PROTEIN PHOSPHORYLATION

Edited by **Claude Prigent**

Protein Phosphorylation
http://dx.doi.org/10.5772/65615
Edited by Claude Prigent

Contributors

Yoshinori Takei, Yoko Amagase, Kiyotsugu Yoshida, Katsuhiko Aoki, Claude Prigent, Janine Post, Stefano Schivo, Indrani Datta, Kavina Ganapathy, Ken-Ichi Yano, Keiko Morotomi-Yano, Sandrine Ruchaud, Omid Feizbakhsh, Matthieu Place, Xavier Fant, Frédéric Buron, Sylvain Routier, Marc Tramier, Florian Sizaire, Catherine Jessus, Aude Dupré

Published by InTech

Janeza Trdine 9, 51000 Rijeka, Croatia

Publishing Process Manager Romina Rovan
Technical Editor SPi Global
Cover InTech Design team

First published November, 2017
Printed in Croatia
Legal deposit, Croatia: National and University Library in Zagreb

Additional hard copies can be obtained from orders@intechopen.com

Protein Phosphorylation, Edited by Claude Prigent
p. cm.
Print ISBN 978-953-51-3625-5
Online ISBN 978-953-51-3626-2

For EU product safety concerns:
IN TECH d.o.o., Prolaz Marije Krucifikse Kozulić 3, 51000 Rijeka, Croatia,
info@intechopen.com or visit our website at intechopen.com.

Meet the editor

Claude Prigent received his PhD degree in Biology at the University in Rennes 1, France, studying DNA ligases in the laboratory of Pr. Michel Philippe. He then worked on DNA repair mechanisms at the Imperial Cancer Research Fund laboratories at South Mimms, London, England, in the laboratory of Pr. Thomas Lindahl. He got a permanent position in the CNRS (French National Center for Scientific Research) just after his postdoc in 1993. He established his own research group in 1996 in Rennes with the main objective on studying the regulation of cell cycle progression by protein kinases with a particular focus on mitosis. His team is also deeply engaged in cancer research trying to understand how deregulated kinases participate in carcinogenesis.

Contents

Preface

Protein phosphorylation reactions are carried out in a cell by protein kinases, which predominantly use ATP as a phosphate donor that is transferred and covalently bound to an amino acid on a substrate protein. Generally, the targeted amino acid is a serine, threonine, or tyrosine. Protein kinases are called serine or threonine kinase because their target is a serine or threonine or called tyrosine kinase because they target tyrosine. Some targeting the three amino acids are called dual-specificity kinases. Protein phosphorylation was discovered in 1954 by Edmond Fischer who shared the Nobel Prize in Medicine or Physiology in 1992 with Edwin Krebs. There are around 500 genes in the human genome, which, thanks to alternative splicing, can generate few thousand protein kinases. There are so many kinases that some have even been called "Just Another Kinase" for JAK kinase. Protein kinases participate in signaling pathways that send a message to a particular place in the cell in particular conditions, just like electric circuits sending electricity to a particular room in a building.

The counterpart of protein kinases is protein phosphatases that remove phosphates from phosphorylated amino acids on a substrate. Protein kinases and phosphatases act as switches in the cell that activate or inactivate protein functions. These enzymatic reactions are reversible, meaning that the cell can quickly react to a particular situation but can also go back to its initial state. For instance, phosphorylation reaches a maximum in mitosis, the shortest cell cycle stage, although it's only a correlation the cell seems to have chosen phosphorylation reaction when it has to control fast biological mechanisms (first section of the book). Considering the importance of phosphorylation in the life of eukaryotic organism, it is thus not surprising that it is found to be at the origin of several diseases (second section of the book). Because identifying the multiple roles of a kinase in a cell or identifying the multiple phosphorylation reactions associated with a biological mechanism remains a challenge, new methodologies and technologies need to be continuously developed (third section of the book).

This book is divided into three sections: "Phosphorylation and Mitosis", "Kinases and Diseases" and "Tools to Study Kinases". The details of the manuscripts that make up these sections are presented below.

The first section "Phosphorylation and Mitosis" is dedicated to the mitotic substrate Arpp19 and to the two protein kinases Haspin and Aurora. **Arpp19** is a protein phosphatase inhibitor, which, when phosphorylated by the kinase Greatwall, specifically inhibits the phosphatase PP2A leaving cdk1/cyclin B substrates phosphorylated to fulfill their function during mitosis. In this review, Dupré and Jessus report that Arpp19 is also phosphorylated by PKA leading this time to an inhibition of CDK1 activation in G2. This dual role of Arpp19 is discussed in the manuscript. **Haspin** is a mitotic histone H3-specific protein kinase that targets threonine 3 in the tail of the histone. This phosphorylation is required for Aurora B localization at kinetochores during mitosis. The structure of Haspin makes it an atypical kinase. Feizbakhsh and collaborators approach all the aspects of structure, regulation, and function of Haspin during the mitosis. **Aurora** kinase family comprises three members in mammals: aurora A, B, and C. Vaufrey and collaborators describe how the kinases are regulated through binding to various activators depending on where and when they need to be active. Regulation through posttranslational modifications is also described.

The second section "Kinases and Diseases" is dedicated to phosphorylations implicated in human pathologies such as cancer and neuronal diseases such as Parkinson's disease.

Priming phosphorylation is a phosphorylation reaction ensured by a priming kinase that transforms a protein into a substrate for a second kinase, known as a processive kinase. Aoki and Yoshika describe several cases of priming/processive kinases and how dysregulation of this mechanism can be involved in cancer. **Nanosecond pulsed electric field** (nsPEF), which is used to treat some solid tumors, triggers nonthermal cell death. In their review, Yano and Morotomi-Yano describe how nsPEF acts as a cellular stress that activates and induces phosphorylation of translation initiation factors and suppression of general protein synthesis. **Parkinson's disease** is characterized by an accumulation of α-synuclein and ubiquitin in cytoplasmic inclusions. Phosphorylation of α-synuclein on serine 129 plays an important role in the disease condition. Datta and Ganapathy report and describe how this phosphorylation event contributes to the disease. The feasibility of **regeneration** to repair damage to the central nervous system increases with time even if the progress remains slow. Phosphorylation of ectodomains of membrane proteins might be a way to trigger regeneration. Takei and Amagase approach these problems in their review.

The third and last section "Tools to Study Kinases" is dedicated to bioinformatic methodology and biosensor technology. The manuscript by Shivo and collaborators is dedicated to the treatment of big data. It describes bioinformatic approaches aimed to use computational modeling to study signaling pathways involved in complex networks. The last manuscript of the book by Sizaire and Tramier is dedicated to biosensors, a very convenient tool used more and more to study protein kinase in vivo. Thanks to the development of new microscopy technologies together with the development of multiple probes corresponding to different fluorescent peptides, there are now different ways to build and use a FRET or FLIM biosensor that are described in this review.

There are so many topics in the field of protein phosphorylation research that this book describes only a tiny part of the domain. I hope that the readers will find useful information in the selected manuscripts. I would like to thank the authors for their contributions and the editors for their work.

Dr. Claude Prigent
Institute of Genetics and Development of Rennes,
CNRS and University of Rennes 1,
Rennes, France

Section 1

Phosphorylation and Mitosis

Chapter 1

ARPP19 Phosphorylations by PKA and Greatwall: The Yin and the Yang of the Cell Decision to Divide

Aude Dupré and Catherine Jessus

Additional information is available at the end of the chapter

http://dx.doi.org/10.5772/intechopen.71332

Abstract

Entry into mitosis and meiosis is orchestrated by the phosphorylation of thousands of mitotic substrates under the control of active Cdk1-cyclin B complexes. To avoid futile cycles of phosphorylation/dephosphorylation, the specific Cdk1-antagonizing phosphatase, PP2A-B55δ, must be simultaneously inactivated. This process is achieved by the activation of the kinase Greatwall (Gwl), which phosphorylates ARPP19. Gwl-phosphorylated ARPP19 then inactivates PP2A-B55δ to allow Cdk1 activation as well as to secure the phosphorylation state of mitotic substrates. This chapter discusses a series of recent works showing that ARPP19 is also phosphorylated by another kinase, PKA. Phosphorylated by PKA, ARPP19 arrests *Xenopus* oocytes in G2 before the first meiotic division. Therefore, depending on its phosphorylation state by either PKA or Gwl, ARPP19 either restrains or activates Cdk1 in *Xenopus* oocytes. Beyond the understanding of the mechanisms of meiotic and mitotic cell division, the control of ARPP19 by its dual phosphorylation enlightens the cAMP-regulated signalization pathways that control vital functions in numerous eukaryotic cell types.

Keywords: cell division, meiosis, ARPP19, Cdk1, PKA, Greatwall, phosphatases, PP2A, kinases

1. Introduction

Cell division is fundamental for life. This process allows the development, the growth and the renewal of any organisms on earth, from bacteria to metazoan. Since its first description by Walther Flemming in 1882, this process has fascinated biologists. M-phase entry and exit are switch-like transitions that are predominantly orchestrated by phosphorylation and dephosphorylation of hundreds of proteins. These "mitotic substrates" govern the structural events that underlie the mechanics of cell division. These phosphorylations are controlled by a cohort of mitotic kinases and phosphatases themselves regulated by two master enzymes: the universal mitotic kinase, Cdk1-cyclin B (or MPF for M-phase promoting factor) and its recently identified antagonizing phosphatase, PP2A, associated to its regulatory subunit B55δ. While we now have a detailed description of cell division mechanics and a long list, still expanding, of mitotic proteins orchestrating these structural cellular events, our basic knowledge of the molecular cascades converging to Cdk1 activation and entry into cell division remains incomplete. In particular, despite the well-documented role of protein kinases in mitosis, little is known regarding the control of Ser/Thr phosphatases that remove the mitotic phosphates and counterbalance the activation and the activity of Cdk1. In 2010, it was shown that the small protein ARPP19 is a substrate of the Greatwall kinase (Gwl) and plays a central role in M-phase through PP2A inhibition. Interestingly, ARPP19 stands for "cAMP-regulated phosphoprotein" and is phosphorylated by PKA in mammalian tissues and cell cultures. In 2014, we showed that ARPP19 phosphorylation by PKA in oocytes maintains these germ cells arrested in G2 by preventing the molecular cascade activating Cdk1. Therefore, ARPP19 is at a crossroad in the meiotic M-phase control network, integrating PKA signaling to halt cell cycle in G2 and switched by Gwl to an M-phase inducer by linking Cdk1 activation to PP2A-B55δ. In this review, we describe how the small ARPP19 protein is able to act as a switcher to guide the cell decision to divide or not. Beyond the understanding of the mechanisms of meiotic and mitotic cell division, deciphering the control of ARPP19 by phosphorylation should also enlighten the cAMP-regulated signalization pathway that controls vital functions in numerous eukaryotic cell types.

2. Cell division: good guys-bad guys, protein kinases-protein phosphatases

2.1. A prominent role of protein kinases ensuring the phosphorylation of mitotic proteins

During cell division, the structure of the cell is dramatically reorganized to properly segregate chromosomes. Since transcription is repressed during cell division, these structural rearrangements are mainly driven by posttranslational modifications of mitotic proteins. Of these, protein phosphorylation is the most abundant event that plays a key role in regulating cell cycle engines. During mitosis entry, approximately 7000 proteins are phosphorylated at multiple sites. This extensive network of phosphorylations is governed by multiple kinases (**Figure 1**), which are organized in cascade, hence amplifying a small initial signal to ensure irreversible transitions during the cell cycle. Of these, the kinase Cdk1 (cyclin-dependent kinase 1) is the most prominent one and is the universal inducer of M-phase entry in all eukaryotes [1–8]. To be active, Cdk1 must be associated to its regulatory subunit, cyclin B (Cdk1-cyclins B complexes, also called MPF for M-phase promoting factor), phosphorylated at the activatory T161 residue and further dephosphorylated at T14 and Y15, two residues that impair its activity when phosphorylated. During M-phase entry, Cdk1 is activated and acts in concert with other kinases to phosphorylate mitotic proteins responsible for the structural changes associated with cell division (**Figure 1**). Hence, entry and progression in mitosis have long been viewed as depending on a giant burst of protein phosphorylation under the control of a master kinase, Cdk1.

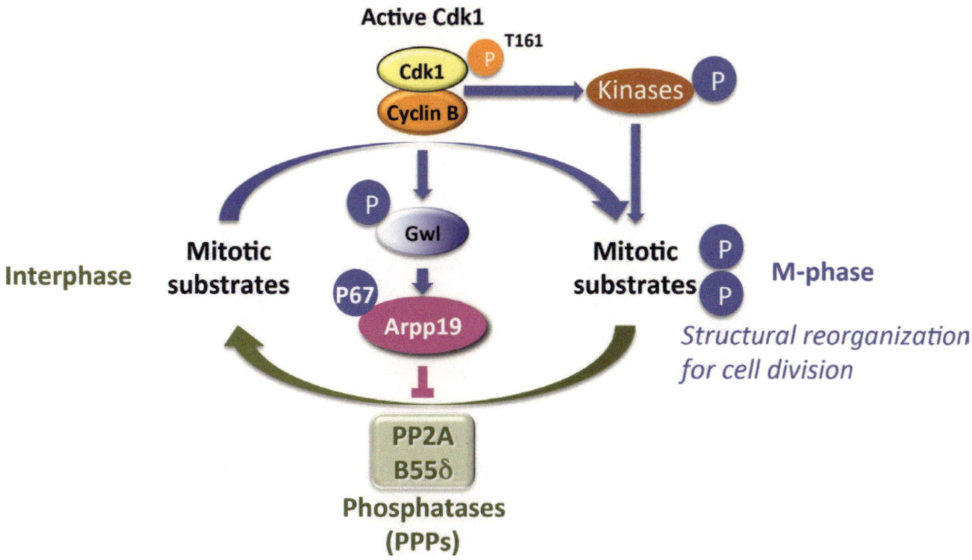

Figure 1. Cell division is controlled by kinases and phosphatases.

2.2. Cell division is not only a matter of kinases: requirement for a fine-tuning of protein phosphatases

Importantly, the return in interphase requires not only Cdk1 inactivation but also the mitotic proteins to be dephosphorylated. This process was at first thought to be controlled by the cellular machinery responsible for protein degradation, as it irreversibly inactivates Cdk1 through cyclin B degradation and consequently leads to the disappearance of phosphorylated mitotic proteins [9, 10]. Nevertheless, mitotic substrates need the action of phosphatases to be dephosphorylated and it is now admitted that the regulation of phosphatase activities is as essential as kinases, not only to exit from M-phase but also for the accurate progression in M-phase (**Figure 1**). Phosphatases are classified into three distinct groups based on their specificity toward residues that are dephosphorylated: the tyrosine phosphatases (cTPs), the dual specificity phosphatases (DSPs) and the serine/threonine phosphatases (STPs) [11]. From the cTPs and DSPs family, only Cdc14, Cdc25 and Ssu72 play an essential role in mitosis [12, 13], and most of the dephosphorylations during M-phase rely on STPs activities. Of these, the phosphoprotein phosphatase (PPPs) family, which includes the protein phosphatases 1 (PP1) and 2A (PP2A), is the largest group necessary for the proper regulation of mitosis [12, 14, 15]. The regulation and the diversity of these phosphatases result from the combinatorial association between conserved catalytic subunits and regulatory subunits that drive their substrate specificity, their cellular localization and their activity. In particular, PP2A is a trimeric complex composed of one catalytic C subunit ($PP2A_C$), one scaffold A subunit ($PP2A_A$) and one regulatory B subunit ($PP2A_B$) that provides substrate specificity. Each subunit is expressed as multiple cellular isoforms, resulting in approximately 100 heterotrimeric PP2A holoenzymes broad-acting *in vivo*. This variety also applies to other PPPs such as PP1, which associates to one or two variable regulatory subunits, thereby creating more than 650 distinct complexes

[14]. Hence, investigating the specific function of these PPPs *in vivo* has been challenging as most of their pharmacological inhibitors, such as okadaic acid or microcystin, do not distinguish among holocomplexes with identical catalytic subunits, yet harboring very different functions due to the variety of their regulatory subunits. Moreover, the effects of these inhibitors have been mostly characterized on purified phosphatases. These phosphatases have been long viewed as housekeeping enzymes, being constitutively active during cell cycle progression. However, failure to regulate their activities has severe consequences on cell division. Inhibiting their activities allows cells to enter mitosis without Cdk1 activation. Then cells proceed through aberrant mitosis with multiple defects on chromosome condensation and segregation [16, 17]. Moreover, their inhibition at metaphase prevents cells to exit from mitosis despite cyclin B degradation and Cdk1 inactivation [18]. More recently, it appeared that it is not just one but a network of PPP activities that temporally orchestrates the dephosphorylation of specific substrates during M-phase. These sequential dephosphorylations order the structural events necessary to exit from cell division, starting with chromosomes segregation and ending with cytokinesis. Hence, by antagonizing kinase activities, the precise regulation of PPP activities is required for the timely execution of mitosis in higher eukaryotes.

In 2010, a breakthrough in the field of cell cycle research came with the discovery that the specific PP2A holoenzyme associated to the B55δ regulatory subunit (PP2A-B55δ) counteracts the Cdk1-dependent phosphorylations in mitosis (**Figure 1**). The activity of PP2A-B55δ is high in interphase and low during mitosis [19]. Like many regulators of the cell cycle, the role and the biochemical mechanism controlling PP2A-B55δ activity were deciphered in CSF and cycling egg extracts from *Xenopus*, a powerful model system that reproduces *in vitro* all the biochemical events of mitosis [20, 21]. Depleting PP2A-B55δ from *Xenopus* egg extracts enhances Cdk1 substrate phosphorylations during mitotic onset and further prevents exit from mitosis due to a failure in dephosphorylating Cdk1 substrates [19]. This was the first evidence that the regulation of a specific phosphatase, PP2A-B55δ, is as important as the Cdk1 kinase for the proper progression in M-phase.

3. Under Gwl control, ARPP19 shuts down PP2A-B55δ to let Cdk1 acting

3.1. ARPP19, a specific inhibitor of PP2A-B55δ

The molecular mechanism underlying the negative control of PP2A-B55δ was further elucidated in *Xenopus* extracts. It revealed that two proteins, the kinase Greatwall (Gwl) and the small protein ARPP19 (cAMP-regulated phosphoprotein of 19 kDa), are the key regulators of PP2A-B55δ. Gwl is an evolutionary conserved protein that was identified in *Drosophila* by two independent genetic screens aimed at identifying new kinases involved in cell cycle progression [22–24]. As cells enter mitosis, Gwl is phosphorylated at multiple sites, probably by various kinases, among them Cdk1 that is critical as it activates Gwl [25–28]. Gwl then phosphorylates two proteins from the endosulfine (ENSA) family: α-endosulfine and its very close paralog ARPP19 at an S residue present in a highly conserved KYFDSGDY motif, corresponding to S67 in *Xenopus* ARPP19 [20, 21, 28]. These small proteins are expressed ubiquitously in neurons as well as in nonneuronal cells and regulate a plethora of biological processes including insulin secretion, quiescence upon nutrient deprivation in yeast and neurite outgrowth [28–34]. Interestingly, these proteins preset the biochemical characteristics of phosphatase inhibitors, being basic-, heat- and acid-stable. When phosphorylated by Gwl at S67, ENSA/ARPP19 is converted into a potent and highly specific inhibitor of PP2A-B55δ that acts by direct binding and does not associate to other PP2A holoenzymes [20, 21] (**Figure 1**). ARPP19 is also a substrate of PP2A-B55δ that binds extremely tightly to the phosphatase and is dephosphorylated at a very low rate [35]. From these results, an "inhibition by unfair competition" model has been proposed wherein phosphorylated ARPP19 would inhibit PP2A-B55δ by blocking access of other substrates [35].

3.2. ARPP19-Gwl, a critical module for the phosphorylation of mitotic substrates, M-phase progression and exit

Because PP2A-B55δ inhibition is required to secure Cdk1 substrate phosphorylation, ENSA/ARPP19 phosphorylation by Gwl is both necessary and sufficient for M-phase entry, progression and exit. Indeed, preventing Gwl inactivation and ARPP19 dephosphorylation abolishes dephosphorylation of mitotic substrates, and consequently M-phase exit despite Cdk1 inactivation [20, 21]. The role of Gwl and ENSA/ARPP19 as key actors governing M-phase progression is largely conserved in somatic eukaryotic cells. In yeast, the Gwl homolog, Rim15p, is required for cellular nutrient responses, such as the arrest of cell proliferation and the induction of meiosis [29, 32, 36]. In human cells, the homolog of Gwl, MASTL, is surprisingly dispensable for the nuclear envelope breakdown (NEBD), a morphological event that signals M-phase entry. However, it remains necessary for important structural events beyond NEBD such as chromosome condensation, proper segregation of chromosomes and cytokinesis [37–43].

The exit from M-phase requires PP2A-B55δ reactivation that depends on Gwl inactivation and ENSA/ARPP19 dephosphorylation. Phosphorylated ARPP19 is present in molar excess over PP2A-B55δ [35], implying that other phosphatases must act during mitosis exit to break the loop established between Gwl, ENSA/ARPP19 and PP2A-B55δ (**Figure 2**). One of these other phosphatases could correspond to PP1. In mitosis, PP1 is inhibited by its Cdk1-dependent phosphorylation at T320. Upon mitosis exit, Cdk1 is inactivated and allows PP1 to dephosphorylate itself at T320 (**Figure 2**). As a result, PP1 is activated and dephosphorylates Gwl, resulting in its inactivation [44, 45]. Consequently, ENSA/ARPP19 is dephosphorylated and no longer inhibits PP2A-B55δ. PP2A-B55δ then dephosphorylates mitotic substrates and the cell shifts from a mitotic to an interphasic structural organization [44, 45]. PP1 also contributes to M-phase exit by dephosphorylating mitotic phosphoproteins [46]. Hence, PP1 acts in concert with PP2A-B55δ for the proper exit from mitosis. In conclusion, ARPP19 is at the center of the antagonistic regulation of an intricate network of kinases and phosphatases, which irreversibly drives M-phase progression by avoiding futile cycles of phosphorylation/dephosphorylation. During the G2/M transition, Cdk1 is activated, whereas PP2A-B55δ is inactivated by ARPP19, thus securing Cdk substrate phosphorylation and progression in M-phase. Upon Cdk1 inactivation by cyclin degradation, both PP1 and PP2A-B55δ are reactivated and consequently dephosphorylate Cdk substrates to promote M-phase exit and the return to interphase.

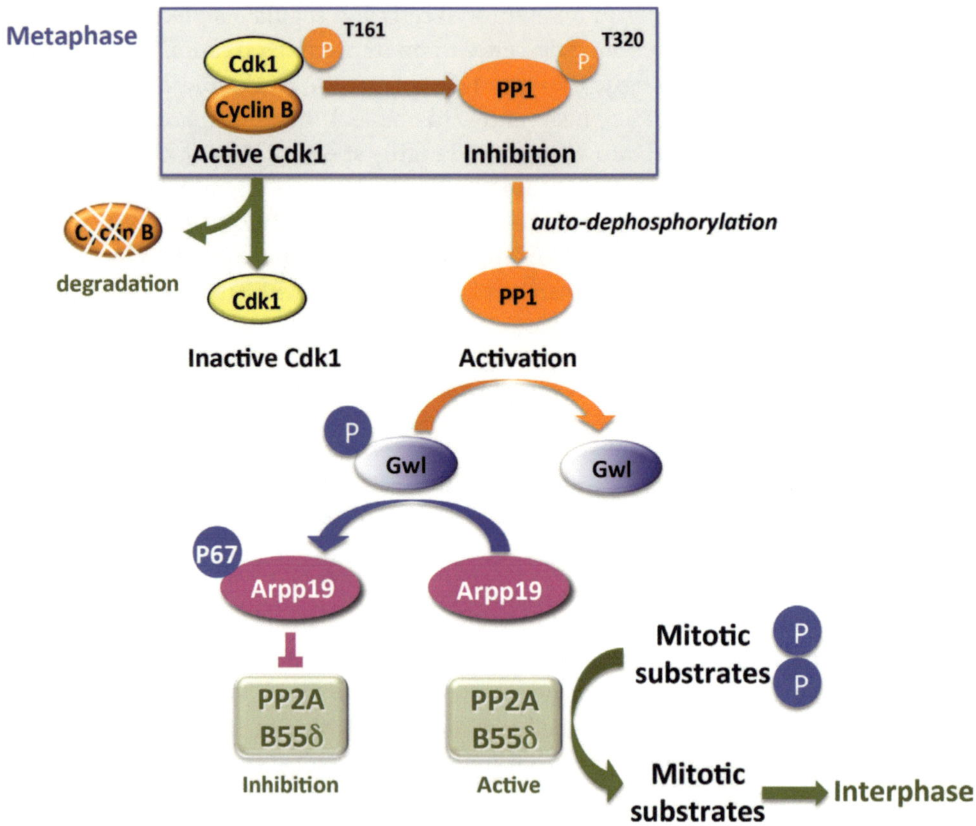

Figure 2. M-phase exit requires PP1 and PP2A-B55δ activation.

3.3. ARPP19-Gwl, a critical module for Cdk1 activation and M-phase entry

Besides acting on Cdk substrate phosphorylation, PPPs are also directly involved in the molecular mechanism responsible for Cdk1 activation. At the end of the G2 phase, Cdk1-cyclin B complexes have accumulated due to cyclin synthesis and are inhibited by Cdk1 phosphorylation at T14 and Y15 (**Figure 3**). These inhibitory phosphorylations depend on the balance of two antagonistic activities: the phosphatase Cdc25 on one side and the Wee1 and/or Myt1 kinases, two redundant members of the same family, on the other side [47–53]. This triad, Wee1/Myt1, Cdc25, and their substrate Cdk1, controls M-phase in all eukaryotes. The activities of Wee1/Myt1 and Cdc25 are regulated by phosphorylation, the first one being inactivated by phosphorylation, whereas the second one is activated. The expression level of Cdk1 being constant during the cell cycle, the formation of Cdk1-cyclin B complexes begins in interphase with the progressive accumulation of cyclin B. As cyclin B accumulates throughout G2, it associates with free Cdk1 molecules that get phosphorylated at T161. These Cdk1-cyclin B complexes are then held inactive by the Wee1/Myt1-dependent phosphorylation of Cdk1 at T14 and Y15 [3, 53]. This prevents precocious entry in M-phase (**Figure 3**). Entry into M-phase is triggered by the activation of Cdc25 that dephosphorylates Cdk1-cyclin B complexes at T14

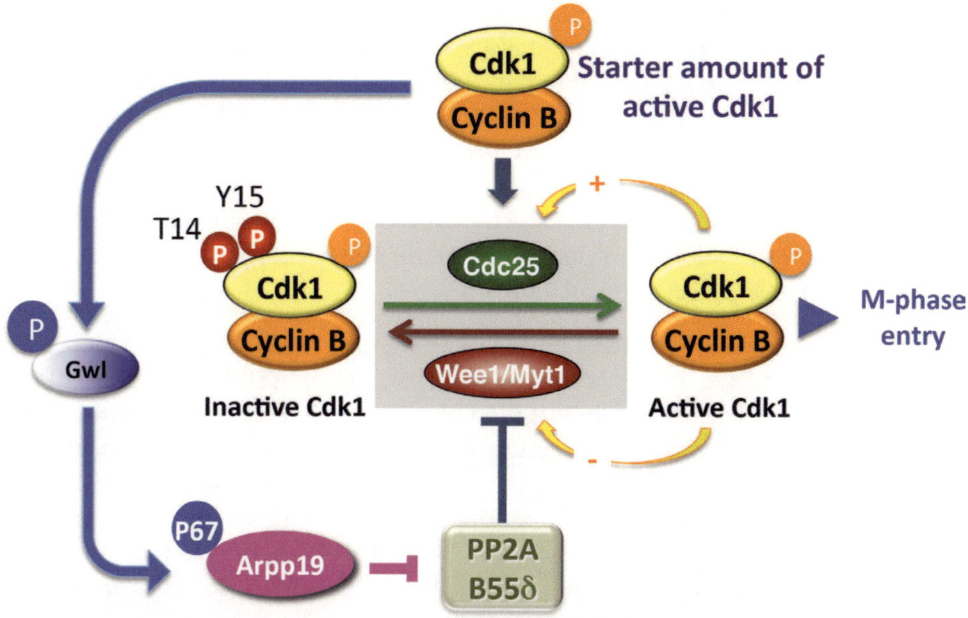

Figure 3. Cdk1 activation and M-phase entry require PP2A-B55δ inactivation.

and Y15. As a result, these complexes are activated and phosphorylate their own regulators, thus enhancing Wee1/Myt1 inhibition and Cdc25 activation [3, 53]: more Cdk1 is activated, more it activates its own activator, Cdc25, and inhibits its own inhibitor, Wee1/Myt1, and more it is activated itself. This mechanism constitutes the core part of a process called the Cdk1 autoamplification loop, which ensures the dephosphorylation of all Cdk1-cyclin B complexes. Furthermore, Cdk1 activates other kinases including Plk1 or Aurora-A to reinforce the loop [3, 54]. Importantly, this switch-like transition is counterbalanced by PP2A-B55δ activity. Indeed, this phosphatase opposes Cdk1 activity by dephosphorylating Cdc25 and Wee1/Myt1, thus compromising the initial activation of Cdk1 by Cdc25 and then the autoamplification process. Importantly, by phosphorylating and activating Gwl, Cdk1 enables ARPP19 phosphorylation at S67 and the consequent inactivation of its counteracting phosphatase. This generates a feedforward regulatory loop to ensure M-phase entry and progression [20–22, 25, 39, 55, 56] (**Figure 3**). Hence, ARPP19, under its Gwl-phosphorylated state, occupies a central position to activate the irreversible switch that engages the cell to divide.

4. PKA, a break of cell division

Kinases are commonly viewed as positive engines driving cell division. Only few kinases, as Wee1/Myt1, restrain the G2/M transition. In G2, some kinases can however inhibit mitosis entry and are usually activated by surveillance mechanisms, termed checkpoints [57]. These kinases, such as ATM, ATR and their substrates Chk1 and Chk2, are activated when DNA integrity is at threat or when DNA replication is not properly completed. By indirectly regulating Wee1/Myt1 and Cdc25 activities, these kinases block M-phase entry and extend the G2 period, allowing the establishment of the DNA quality control and repair processes [57]. However, other kinases have the potential to block M-phase entry without being involved in checkpoint mechanisms. Of these is the well-conserved cAMP-dependent protein kinase A (PKA). PKA is the major mediator of the pleiotropic cellular regulator, cyclic AMP (cAMP). This kinase is one of the most prominent actors of signal transduction pathways and plays pivotal functions in various cellular processes including metabolism, cell survival, proliferation, differentiation and cell cycle regulation.

PKA is a heterotetradimer composed of two catalytic subunits (PKA_C) and two regulatory subunits (PKA_R). The PKA_C subunits contain the active site, an ATP-binding site and a binding domain for PKA_R. The PKA_R subunits bind one another in an antiparallel orientation to form a homodimer and further encompass a cAMP-binding domain also interacting with PKA_C and an "autoinhibitory" domain that serves as a pseudosubstrate for PKA_C. When the intracellular concentration of cAMP is low, PKA_R dimer binds and inactivates PKA_C. As the concentration of cAMP increases, cAMP binds to PKA_R and induces an allosteric conformational change, which unleashes PKA_C subunits. Free PKA_C subunits are then active and phosphorylate their intracellular targets. This process further involves scaffold proteins, termed AKAPs (A-kinase anchoring proteins), which tether the kinase to defined cellular sites in order to confer spatial and temporal specificity to PKA [58, 59]. While most of its contribution involves the activation of gene transcription, nongenomic effects of PKA have been implicated in M-phase progression. Remarkably, high PKA activity is responsible for arresting vertebrate oocytes in prophase of the first meiotic division by indirectly inhibiting Cdk1 activation. This negative effect of PKA exerted on M-phase entry is also conserved in somatic cells, in which keeping high levels of cAMP or PKA activity was found to prevent Cdk1 activation and block the G2/M transition [60–63]. While the role of PKA as a potent inhibitor of the mitotic and the meiotic cell division is well established since almost 40 years, its substrate in both somatic cells and oocytes had remained unidentified all over these years. Using meiotic division of *Xenopus* oocytes as a model system, this critical question has received a first answer in 2014 [64].

4.1. The oocyte meiotic division, a powerful system model to study the G2/M transition

The arrest in prophase is specific to female germ cells and is universal in the animal kingdom. This arrest allows oocytes to accumulate the components necessary to support embryo development. On the contrary to somatic cells, this physiological arrest is independent of checkpoint mechanisms. At the time of ovulation, the prophase arrest is released by external stimuli. In vertebrates, the luteinizing hormone (LH) promotes the synthesis, in ovarian follicular cells, of a second hormonal inducer that acts on the oocyte to induce meiosis resumption. In lower vertebrates as amphibians, this signal corresponds to a steroid hormone, progesterone [65–69]. Upon the hormonal stimulation, oocytes complete the first meiotic division and proceed through the second meiotic division with no intervening S-phase, which reduces the parental genome by half (**Figure 4**). These cells then halt their

http://dx.doi.org/10.5772/intechopen.71332

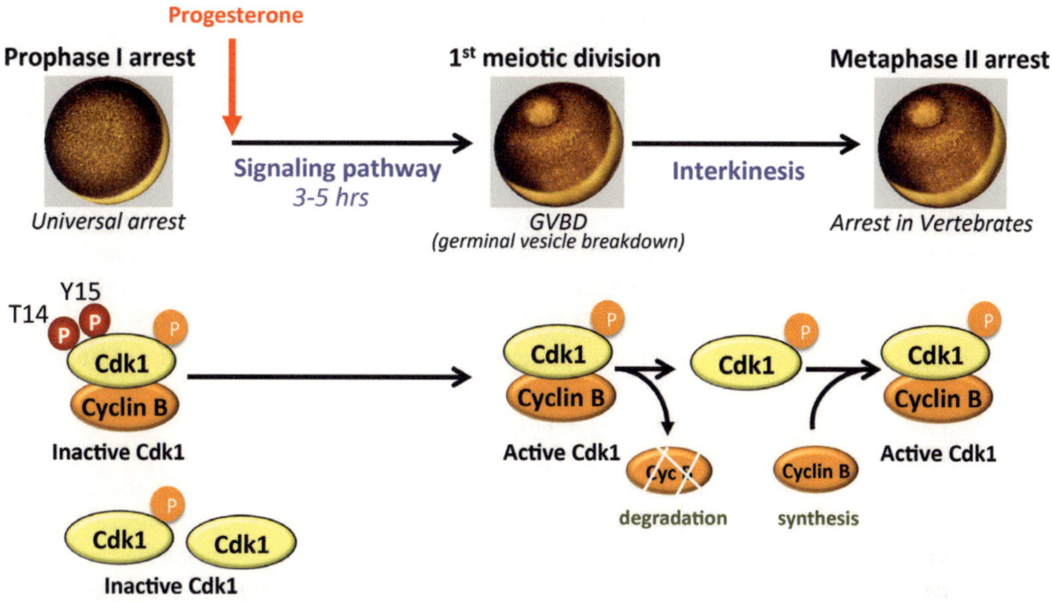

Figure 4. Meiotic maturation of *Xenopus* oocytes.

divisions in metaphase of the second meiotic division (metaphase II) in all vertebrates. This process, called meiotic maturation, generates haploid gametes ready to be fertilized. Similarly to mitosis, the progression through the meiotic maturation depends on Cdk1 that sustains two successive waves of activation to coordinate the two consecutive cell divisions (**Figure 4**). The first activation of Cdk1 promotes the reentry in the first meiotic division, a process marked by the nuclear envelope breakdown, or germinal vesicle breakdown (GVBD). Thereafter, Cdk1 is inactivated in metaphase I by the ubiquitin-dependent degradation of cyclin B to induce exit from the first meiotic division and the extrusion of the first polar body [70]. Concomitantly to cyclin B degradation, newly synthesized cyclin B molecules reaccumulate and quickly reactivate Cdk1 to enable entry into the second meiotic division (**Figure 4**). In metaphase II, an oocyte-specific kinase network, the Mos/MAPK/Rsk module, and its substrate Erp1/Emi2 inhibit cyclin B degradation [71–74]. This process stabilizes Cdk1 activity at a high level and arrests the cell in metaphase II, a specific feature of vertebrates.

The meiotic cell division shares most of the molecular regulators with mitosis. The components of the Cdk1 autoamplification loop as well as Gwl, ARPP19 and various PPPs such as PP1 and PP2A are conserved. Hence, the arrest in prophase of oocytes is commonly assimilated to a late G2 phase and the process of meiosis resumption has served as an historical model to decipher signaling pathways involved in the G2/M transition. Of these, *Xenopus* oocyte has been widely studied to decrypt the biochemical regulation of cell division. In this species, progesterone induces Cdk1 activation within 3 to 5 hours, a process visualized by GVBD (**Figure 4**). Hence, it is possible to investigate meiosis resumption from the first molecular events induced by the hormone until Cdk1 activation.

4.2. PKA, acting both at the top and at the end of the signaling pathway leading to Cdk1 activation

Xenopus prophase–arrested oocytes contain two distinct pools of inactive Cdk1: a monomeric form of Cdk1 that is already phosphorylated at T161 and a stockpile of Cdk1-cyclin B complexes phosphorylated at T14 and Y15 [70, 75, 76] (**Figure 4**). These complexes are continuously formed during the prophase arrest because of the slow accumulation of cyclin B that associates with free T161-phosphorylated Cdk1 [70, 75]. As in somatic cells, these newly formed Cdk1-cyclin B complexes are immediately phosphorylated by Myt1 at T14 and Y15 and inactivated [77–79]. Cdc25 is kept inactive, likely resulting from its S287 phosphorylation by PKA that promotes its cytoplasmic retention and from the dephosphorylated state of its Cdk1 residues [80].

4.2.1. Initiation of the signaling pathway: a PKA lock

In all vertebrates, PKA activity is responsible for the prophase arrest of oocytes. In these resting cells, PKA is kept active by elevated intracellular levels of cAMP, necessary to restrain Cdk1 activation [81–87]. In *Xenopus* oocytes, although the receptor of progesterone has not been identified, the earliest event detectable within 10 minutes upon the hormonal stimulation is a decrease of cytoplasmic cAMP concentration, immediately followed by the downregulation of PKA activity (**Figure 5**). The drop in cAMP concentration and PKA downregulation are both necessary and sufficient for unlocking a molecular cascade that activates Cdk1 3–5 h later. Forskolin, cholera toxin and IBMX, three cAMP-elevating agents, or recombinant PKA_C, are all potent inhibitors of progesterone-induced meiosis resumption [88–91]. Conversely, injection of recombinant PKA_R or a specific protein inhibitor of PKA, PKI, in the oocyte is sufficient to activate Cdk1 in the absence of progesterone [92, 93]. Hence, both cAMP and PKA, by keeping Cdk1 inactive, are the key actors of the arrest of oocytes in prophase. This implies that a substrate of PKA is responsible for the arrest in prophase under its phosphorylated form (**Figure 5**). Likewise, this PKA substrate must be dephosphorylated within 1 h after hormonal stimulation to launch the signaling pathway that ultimately authorizes Cdk1 activation 3–5 h later.

4.2.2. The core of the signaling pathway: cyclin B and Mos accumulation

The few hours long pathway connecting PKA inhibition and Cdk1 activation is not fully known. PKA inhibition controls the synthesis of new proteins from a preexisting pool of maternal mRNAs in *Xenopus* oocyte. Two newly synthesized proteins are essential for Cdk1 activation: cyclin B, in particular the B1 isoform, and the kinase Mos (**Figure 5**). Mos is a germ cell–specific kinase. It activates MEK that in turn activates MAP kinase (MAPK/ERK) that phosphorylates, among other substrates, $p90^{Rsk}$. Mos starts to accumulate to detectable levels at GVBD and activates the Mos/MAPK pathway during the whole process of meiotic maturation in all animal species analyzed so far [94–96]. Mos and cyclin B1 are both synthesized in response to progesterone or following PKI injection [94, 97]. In contrast, they do not accumulate in response to progesterone if PKA is experimentally kept active [97], hence demonstrating that their accumulation, essential in the pathway converging to Cdk1 activation, is under the negative control of PKA.

Cyclin B and Mos activate two redundant pathways ending with the activation of Cdk1. While suppressing by an antisense strategy the synthesis of one pathway, either cyclin B or

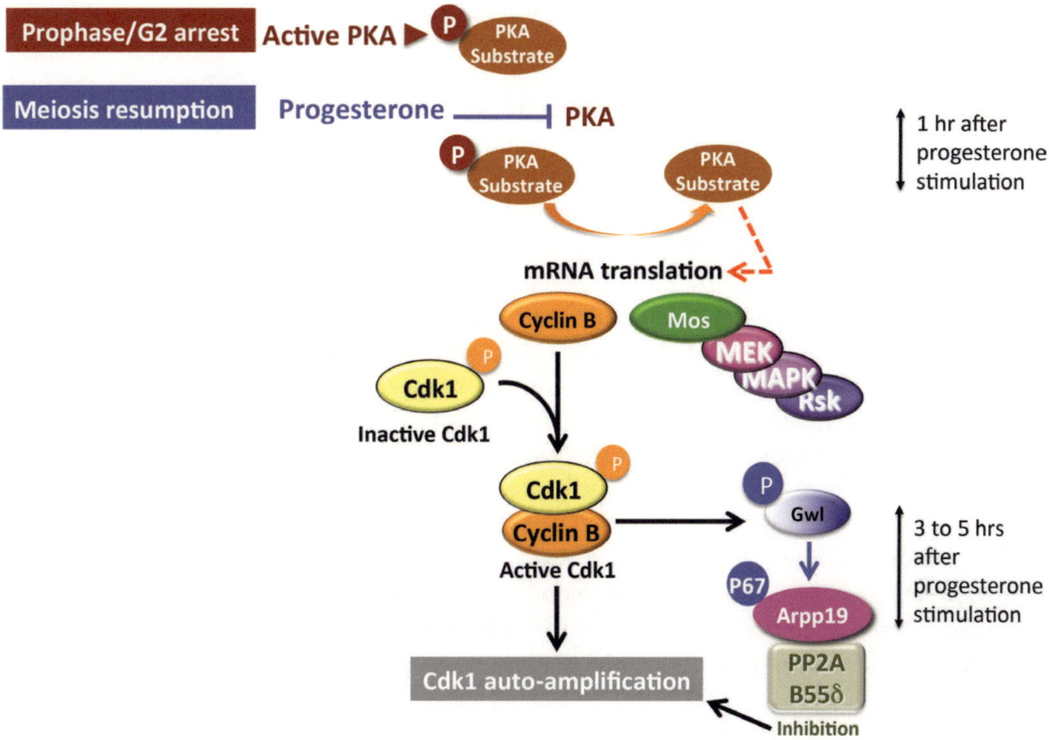

Figure 5. The signaling pathway leading to Cdk1 activation in *Xenopus* oocytes.

Mos, does not hinder meiosis resumption upon progesterone stimulation or PKI injection in *Xenopus* oocyte, inactivating simultaneously both pathways completely abolishes meiosis resumption [97, 98]. In this double knock-down condition, Cdk1 activation is restored by the replenishment of either one of them, provided by injecting oocyte with either recombinant Mos or cyclin B [97, 98]. While both proteins contribute to Cdk1 activation, cyclin B is likely the principal actor required for this process. In response to progesterone, the rate of cyclin B1 synthesis and its accumulation increase independently of Cdk1 activation [99]. On the contrary, Mos accumulates only if the protein has been primarily stabilized by its Cdk1-dependent phosphorylation [99, 100], therefore suggesting that its accumulation requires basal levels of Cdk1 activity that are not detectable in prophase oocytes. The accumulation of cyclin B1 molecules generates new Cdk1-cyclin B complexes from monomeric T161-phosphorylated Cdk1 [77, 99, 101] (**Figure 5**). Accordingly, cyclin B1 or B2 injection promotes Cdk1 activation independently of protein synthesis [97, 98], whereas overexpressing a dominant negative form of Cdk1, K33R-Cdk1, known to sequester newly synthesized cyclins, prevents progesterone-induced meiosis resumption [101]. The newly formed Cdk1-cyclin B complexes escape the Myt1-dependent inhibitory phosphorylations of Cdk1 at T14 and Y15 by an unknown mechanism [76, 77]. Hence, these newly formed Cdk1-cyclin B complexes are immediately active [77]. In this model, the Mos/MAPK pathway would contribute to Cdk1 activation by phosphorylating Myt1 and Cdc25, hence promoting their respective inhibition and activation [94, 102].

4.2.3. The endpoint of the signaling pathway: the Cdk1 autoamplification loop, a process sensitive to PKA

The few active molecules of Cdk1 generated by cyclin B1 accumulation serve as a trigger to launch the Cdk1 autoamplification loop that converts the stockpile of inactive Cdk1-cyclin B complexes into active ones in an autoaccelerated manner. This process relies entirely on the dephosphorylation of inactive Cdk1-cyclin B complexes at T14 and Y15 (**Figure 6**). As in mitosis, this process depends primarily on Cdk1 activity that phosphorylates its own regulators, Cdc25 and Myt1, resulting in their respective activation and inhibition. It is reinforced by other kinases such as the Mos/MAPK pathway, Plk1 or Aurora A, which are activated downstream of active Cdk1 and also target Cdc25 and Myt1. Importantly, PPP phosphatases that counterbalance Cdk1 activity must be inhibited as well, to avoid the dephosphorylation of Cdc25 and Myt1 (**Figure 6**). Indeed, inhibiting both PP1 and PP2A directly triggers the Cdk1 autoamplification loop independently of the upstream steps of the signaling pathway induced by progesterone, i.e. PKA downregulation and protein synthesis, showing that PPPs are part of the core machinery of the autoamplification loop [17, 103, 104]. Moreover, PP2A-B55δ inactivation is likely the key PPP involved in this process as shown by the important role of Gwl in this process: Gwl is activated upon Cdk1 activation in oocytes of a range of species (starfish, *Xenopus* and mice) (**Figures 5** and **6**) and the overexpression of a constitutive active form of Gwl is sufficient to activate Cdk1 in *Xenopus* oocytes [105, 106].

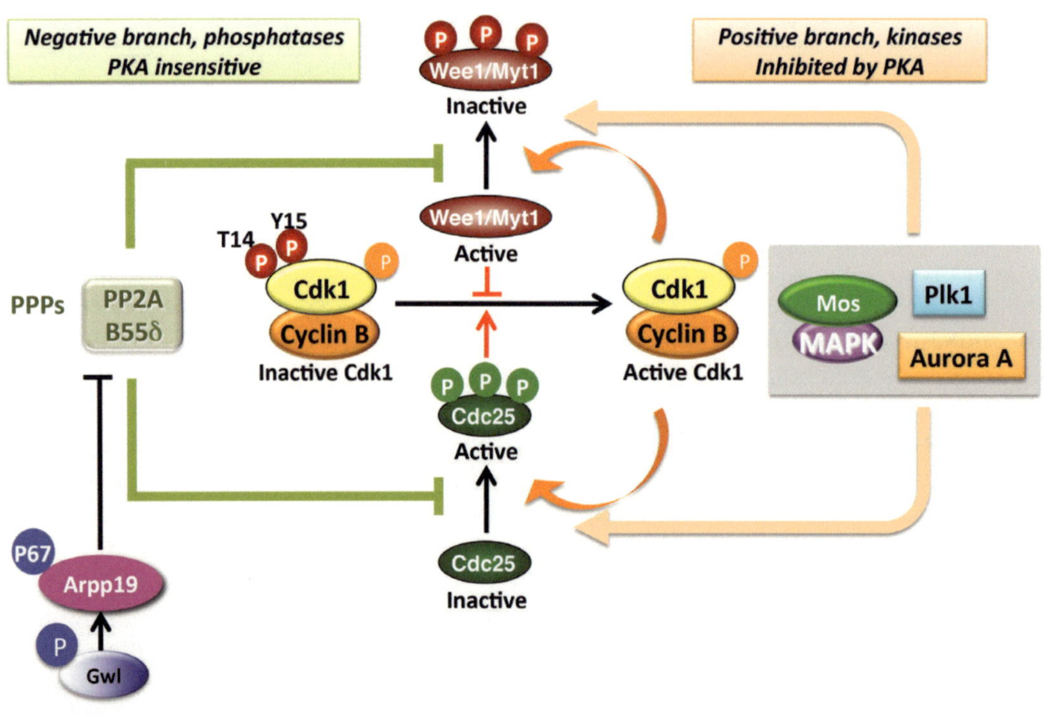

Figure 6. The Cdk1 autoamplification loop.

Consequently, the autoamplification loop can be directly activated in the oocytes using two experimental ways that bypass the upstream signaling cascade: either by inhibiting the negative branch of the loop, i.e. PPPs, by example with okadaic acid or microcystin; or by overexpressing the actors of the positive branch of the loop, such as cyclin B, Mos or Cdc25. It has to be noted that PKA is not active when the loop takes place [107, 108]. Indeed, PKA is inhibited in response to progesterone within less than 1 h and stays at a low activity level thereafter. Strikingly, an artificially elevated activity of PKA prevents all the positive actors of the Cdk1 autoamplification loop, as purified Cdk1-cyclin B complexes, cyclin B, Mos or Cdc25 to launch the loop [70, 103, 105, 109, 110]. In contrast, PKA activity does not prevent the autoamplification loop when triggered by the inhibition of PPPs [17, 104]. Thus, PKA activity is unable to control PPP activities involved in the loop but has the ability to block the positive actors of the loop by targeting unidentified substrates among them. Importantly, the control of the autoamplification loop exerted by PKA is not physiological, as this kinase is catalytically inactive at the time of Cdk1 activation [107, 108]. Thus, PKA can be pictured as a checkpoint mechanism operating under extraphysiological conditions: any abnormal elevation of PKA activity at that time will block meiotic division.

Hence, understanding the signaling pathway that triggers the G2/M transition in *Xenopus* oocytes requires the identification of the PKA substrate(s) that account for the inhibitory function of PKA at the very beginning of the signaling pathway as well as at the endpoint of this pathway, the autoamplification loop. We have undertaken this quest and found one single protein answering these issues. To our great surprise, this protein corresponds to ARPP19.

5. Under PKA control, ARPP19 is at the core of the cell decision-making: to divide or not

5.1. In search of candidate proteins corresponding to the long-sought oocyte PKA substrate

Since the 1980s, multiple attempts have been unsuccessfully undertaken to identify the substrate of PKA that blocks the initiation of the signaling cascade leading to Cdk1 activation in oocytes. In 2002, Cdc25 was shown to be phosphorylated at S287 by PKA, a process contributing to its inactivation in *Xenopus* prophase–arrested oocytes [80]. Upon progesterone stimulation, Cdc25 is dephosphorylated at S287 and phosphorylated at multiple activatory residues under the control of Cdk1, which activates its catalytic phosphatase activity and allows Cdk1 activation [80, 111]. However, Cdc25 does not fulfill the expected biochemical criteria for being the PKA substrate whose dephosphorylation triggers Cdk1 activation. In particular, S287 dephosphorylation of Cdc25 relies on protein synthesis and occurs much later than PKA inhibition, i.e. at the time of Cdk1 activation [80]. More importantly, S287 dephosphorylation depends on Cdk1 activity and therefore cannot represent the upstream event accounting for Cdk1 activation [80]. Another candidate is the kinase Wee1. In mice, Wee1 is expressed in prophase oocyte and its phosphorylation by PKA enhances its inhibitory effect toward meiosis resumption [112, 113]. However, the only kinase targeting Cdk1 at T14 and Y15 that is expressed in *Xenopus* oocytes is Myt1, Wee1 being only expressed after the completion of the first meiotic division [114]. Thus, Wee1 cannot be the PKA substrate responsible for the prophase arrest in this species.

Going back in the 1980s, a biochemical approach using *Xenopus* prophase–arrested oocytes led to the purification of a heat- and acid-stable protein of 20 kDa, described as a PKA substrate and dephosphorylated 1 h after progesterone stimulation, downstream PKA inhibition [115]. The close similarities of the biochemical properties of this protein (heat- and acid-stability, molecular weight) with ARPP19 have drawn our attention. While the cell cycle field was focused on Gwl-ARPP19, a comeback to the origins of the ARPP19 discovery, hidden inside its fully developed name cAMP-regulated phosphoprotein 19, highlights that ARPP19 is a substrate of PKA, phosphorylated at an S residue within a PKA consensus RKP/SSLV motif conserved among most animals [21, 64, 116, 117]. Indeed, the acid-soluble proteins of the ENSA family, to which belongs ARPP19, were originally identified as a group of phosphoproteins upon dopaminergic stimulation [117–119]. The signaling pathway induced by dopamine involves cAMP production and PKA activation. Various cellular functions have been attributed to the PKA phosphorylation of these family members [118, 119]. Of these, DARPP32, a member of this family, becomes a potent inhibitor of PP1 when phosphorylated by PKA [118, 119]. ENSA/ARPP19 interacts selectively with the membrane-bound form of the presynaptic protein α-synuclein [120]. When constitutively phosphorylated by PKA, the interaction between ENSA/ARPP19 and α-synuclein is abolished, interfering with α-synuclein self-assembly in membrane, and probably accounting for the role of ENSA/ARPP19 in neurodegenerative diseases [120]. Nevertheless, no specific function related to cell division had been attributed to this particular PKA phosphorylation until we hypothesized that ARPP19 could be the substrate of PKA in prophase-arrested oocytes.

5.2. The phosphorylation of ARPP19 at S109 by PKA arrests oocytes in G2

The *Xenopus* ARPP19 sequence encompasses the residue targeted by PKA in the previous studies conducted on the ENSA family [21, 64, 117]. This residue corresponds to S109 in *Xenopus* and is part of a PKA consensus sequence (RKPS_{109}LVA) well conserved among eukaryotes. At first glance, it appears provocative to propose ARPP19, a protein recently shown to be essential for Cdk1 activation, as the PKA substrate ensuring the prophase arrest by locking the transduction cascade ending with Cdk1 activation. How to conceive that the same molecule would be negative for M-phase entry and thereafter transformed into a positive actor of cell division? This apparent paradox is circumvented by the fact that these antagonistic functions do not take place at the same period in oocytes (being negative at the top of the cascade and then positive at its endpoint), do not target the same molecular systems (early players of the cascade for the negative function, PP2A-B55δ for the positive function) and are provided by distinct kinases (PKA for the negative function, Gwl for the positive function).

Indeed, we have shown that *in vivo*, ARPP19 is phosphorylated at S109 by PKA in prophase-arrested oocytes. This residue is dephosphorylated within 1 h after progesterone stimulation or in response to PKA inhibition provoked by PKI injection [64]. A phosphomimetic S109 mutant, S109D-ARPP19, proved to be a strong inhibitor of Cdk1 activation when injected into oocytes, not only in response to progesterone but also following PKI, Mos or cyclin B injections. However, S109D-ARPP19 does not impair the Cdk1 autoamplification loop promoted by injecting the phosphatase inhibitor, okadaic acid [64]. These results demonstrate that the PKA-dependent phosphorylation of ARPP19 at S109 restrains Cdk1 activation and is responsible for the prophase-arrest of *Xenopus* oocytes (**Figure 7**). Upon progesterone stimulation, PKA is inhibited and ARPP19 is dephosphorylated at S109. This event is necessary to initiate the signaling cascade that ultimately generates the threshold level of active Cdk1, which consequently triggers the Cdk1 autoamplification loop (**Figure 7**). Importantly, the Cdk1 autoamplification loop does no longer require the dephosphorylation of ARPP19 at S109. ARPP19 is therefore the long-sought substrate of PKA that maintains oocytes in prophase. The identity and the regulation of the phosphatase that dephosphorylates ARPP19 at S109, hence unlocking the signalization cascade converging to Cdk1 activation, as well as the direct molecular targets of phospho-S109 versus dephospho-ARRP19 that form the latch of this cascade, now deserve investigation.

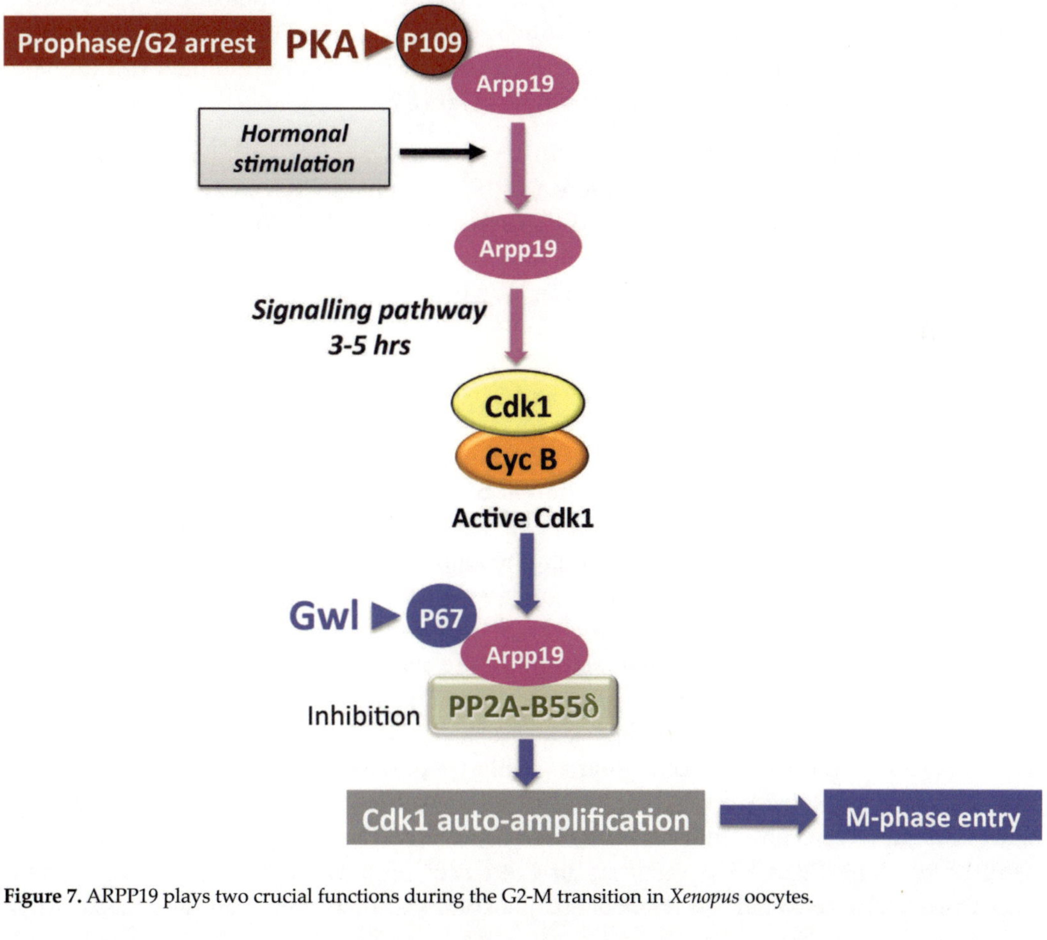

Figure 7. ARPP19 plays two crucial functions during the G2-M transition in *Xenopus* oocytes.

5.3. The phosphorylation of Arpp19 at S67 by Gwl promotes Cdk1 activation and M-phase independently of PKA activity

Strikingly, overexpressing ARPP19 in *Xenopus* oocytes produces opposing effects depending on the amount of the injected protein. High amounts of ARRP19 inhibit meiotic divisions induced by progesterone, whereas decreasing the amount of injected ARPP19 stimulates meiosis resumption [64]. This experiment reveals the antagonistic functions of ARPP19 in oocytes: it inhibits Cdk1 activation when phosphorylated by PKA (as when injected at high concentration) and is converted into a positive regulator of Cdk1 activation during meiosis resumption (as when injected at low concentration). This latter role was then investigated carefully during oocyte meiotic maturation. We showed that ARPP19 is phosphorylated at S67 by Gwl at the time of Cdk1 activation. Once phosphorylated by Gwl, ARPP19 interacts with PP2A-B55δ [110]. We then produced *in vitro* a Gwl-thiophosphorylated form of ARPP19 protein, which is resistant to dephosphorylation. Injecting this constitutively phosphorylated form of ARPP19 at S67 promoted meiosis resumption independently of progesterone and of the synthesis of either Mos or cyclin B [110]. We also generated a nonphosphorylatable ARPP19 mutant at S67 by mutating S67 into an alanine (S67A-ARPP19). The S67A-ARPP19 mutant proved to be a strong inhibitor of meiosis resumption in response to either progesterone or PKI injection but also following the overexpression of constitutive active Gwl [110]. These results clearly demonstrate that ARPP19, once phosphorylated by Gwl at S67, is involved in the Cdk1 autoamplification loop by directly inhibiting PP2A-B55δ (**Figure 7**).

Importantly, the S67-phosphorylated ARPP19 protein was able to interact with PP2A-B55δ and to activate Cdk1 in the presence of high levels of PKA activity [110]. Hence, when phosphorylated by Gwl at S67, ARPP19 inactivates PP2A-B55δ and enables the Cdk1 autoamplification loop to function independently of PKA activity, in strong agreement with the PPPs of the Cdk1 autoamplification loop being insensitive to the inhibitory action of PKA (**Figure 6**).

5.4. The functional interplay between S67 and S109 phosphorylations

When phosphorylated at S109 by PKA, ARPP19 inhibits the molecular cascade leading to Cdk1 activation. Upon Gwl phosphorylation at S67, ARPP19 becomes a potent inducer of Cdk1 activation by inactivating PP2A-B55δ. Hence, ARPP19 acts as a switch in the mechanism of Cdk1 activation depending on its phosphorylation state at S109 and S67. It was then important to determine if both phosphorylations counteract each other, explaining their antagonistic effects. We investigated whether S109 phosphorylation of ARPP19 could prevent its S67 phosphorylation by Gwl, accounting for its inhibitory action toward M-phase entry. *In vitro*, Gwl is able to phosphorylate recombinant ARPP19 at S67 independently of its phosphorylation at S109 [116, 121]. The injection of a double phosphorylated form of ARPP19 at S109 and S67 induces Cdk1 activation as well as meiosis resumption, and the protein is able to interact with PP2A-B55δ [121]. Hence, the phosphorylation of ARPP19 at S109 does neither impair its own phosphorylation by Gwl nor abolish the ability of S67-phosphorylated ARPP19 to inhibit PP2A-B55δ, resulting in Cdk1 activation through the autoamplification loop. Once Gwl is activated, the effect of S67 phosphorylation is dominant over the negative function conferred by the S109 phosphorylation. Accordingly, the phosphomimic S109D-ARPP19 mutant is phosphorylated at S67 upon the overexpression of a constitutive active Gwl or the injection of S67-phosphorylated ARPP19 and becomes unable to block meiosis resumption under these conditions [64]. Likewise, the inhibition of PP2A-B55δ resulting from S67 phosphorylation of ARPP19 activates Cdk1 independently of PKA. Therefore, the PKA-dependent phosphorylation of ARPP19 does not exert its negative control of M-phase by antagonizing the biological ability of S67-phosphorylated ARPP19 to activate Cdk1. Moreover, the effect of S67 phosphorylation is dominant over the negative function of S109 phosphorylation.

Remarkably, ARPP19 is rephosphorylated at S109 at the time of Cdk1 activation, concomitantly with the S67 phosphorylation by Gwl [64]. This rephosphorylation of ARPP19 at S109 is promoted by a distinct kinase than PKA, not yet identified [64] (**Figure 8**). Therefore, in contrast to what was previously thought, it is not a single S67-phosphorylated form of ARPP19 that activates Cdk1, but a double S109-S67 phosphorylated form. We investigated whether S109 rephosphorylation of ARPP19 contributes to the Cdk1 autoamplification loop or is neutral. Injecting an *in vitro* S67-phosphorylated form of ARPP19 that cannot be phosphorylated at S109, the S109A mutant, is less efficient to activate Cdk1 than S67-phosphorylated ARPP19 [121]. These results show that the active form of ARPP19 at the end of the signaling pathway corresponds to a double phosphorylated form of ARPP19 at both S67 and S109, and that S109 phosphorylation confers new properties to ARPP19, contributing to Cdk1 activation in a context where S67 is already phosphorylated (**Figure 8**).

Hence, these results highlight the importance of timely synchronizing ARPP19 phosphorylations at S109 and S67 to properly activate Cdk1 in *Xenopus* oocytes. In prophase, ARPP19 phosphorylation at S109 by PKA restrains Cdk1 activation. Upon hormonal stimulation, ARPP19 is dephosphorylated at S109, which launches the signaling cascade. At its endpoint, ARPP19 is phosphorylated at both sites, S109 by an unknown kinase and S67 by Gwl. It then becomes a positive actor within Cdk1 autoamplification loop (**Figure 8**). In the future, it will be important to identify the kinase responsible for ARPP19 phosphorylation at S109 and to determine how the rephosphorylation of ARPP19 at S109 positively regulates the Cdk1 autoamplification process.

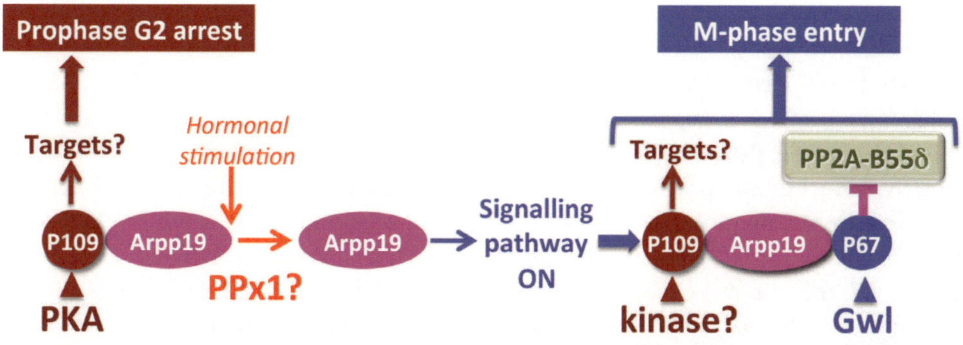

Figure 8. Synchronized phosphorylation of ARPP19 controls the G2-M transition of *Xenopus* oocytes.

The phosphorylations of ENSA/ARPP by PKA and Gwl, acting as a switch-like process to regulate PP2A activity, are found in other cell types, in particular nondividing cells. In postmitotic striatal neurons, ARPP19 is not expressed but its close relatives, ARPP16 and ARPP21, act as potent inhibitors of PP2A. ARPP16 exhibits less specificity toward PP2A isoforms than ARPP19, as it inactivates both PP2A-B55α and PP2A-B56δ [122]. In great contrast to oocytes, in which Gwl is activated as the cell undergoes meiotic divisions, MAST3, the homolog of Gwl, is active and phosphorylates ARPP16 in striatal cells [122]. The constitutive activation of the MAST3-ARPP16 module maintains PP2A continuously inhibited in the absence of stimulation, in order to avoid the dephosphorylation of PP2A substrates [122]. Upon dopaminergic stimulation, PKA is activated and then phosphorylates ARPP16, a process accompanied by its MAST3-dependent dephosphorylation [122]. Importantly, the phosphorylations of ARPP16 by PKA and MAST3 are mutually exclusive and the prior phosphorylation of ARPP16 by PKA reduces the ability of MAST3 to phosphorylate ARPP16 [123]. This reciprocal regulation of ARPP16 by PKA and MAST3 provides a switch in PP2A inactivation, under the control of cAMP-PKA. Hence, in striatal cells, the phosphorylation of ARPP16 by PKA likely attenuates PP2A inactivation promoted by MAST3-phosphorylated ARPP16, thus creating a mechanism whereby cAMP deinhibits PP2A. Therefore, a common set of molecular players, ARPP, PKA and Gwl, is used to regulate PP2A activity in response to various extracellular signals regulating cAMP, but in distinct ways depending on the cell type and/or on dividing or nondividing cells.

6. Concluding remarks

In light of these recent advances, a new model for the regulatory mechanisms controlling the G2/M transition and ending with Cdk1 activation emerged. In this model, two phosphatases would act at different times during meiosis resumption to control the signaling pathway induced by progesterone and ending with the Cdk1 autoamplification loop (**Figure 8**). Upon progesterone stimulation, PKA is downregulated and a first phosphatase (PPx1) dephosphorylates ARPP19 at S109 within 1 h. The role of PPx1 is essential for meiosis resumption as it generates a S109-dephosphorylated form of ARPP19 that licenses the signaling cascade, ending with Cdk1 activation. The intracellular pathway between S109-dephosphorylated Arpp19 and Cdk1 activation is not fully elucidated yet. However, it passes through the accumulation of two proteins: Mos and cyclin B, essential to generate a threshold amount of active Cdk1 in *Xenopus* oocytes [124]. This basal level of Cdk1 activity is not sufficient to orchestrate the structural reorganization needed for the cell to divide. But it certainly activates Gwl, leading to the phosphorylation of ARPP19 at S67 and to the inhibition of a second phosphatase, PP2A-B55δ. Surprisingly, S109 rephosphorylation of ARPP19 enhances the positive effect of S67 phosphorylation on Cdk1 activation (**Figure 8**). Thanks to PP2A-B55δ inhibition and the unknown targets of the S109 phosphorylation of ARPP19, the Cdk1 autoamplification loop becomes functional and enables the irreversible commitment of oocytes in the meiotic cell division. In this new model for Cdk1 activation, the activity of PPx1 unlocks the top of the signaling pathway in response to progesterone, whereas the inhibition of PP2A-B55δ is necessary several hours later for the Cdk1 autoamplification loop responsible for entry into M-phase.

Over the past decade, tremendous advances in our understanding of M-phase progression have been made, in particular regarding the role and the regulation of mitotic phosphatases. The concerted but opposite action of kinases and phosphatases allows for the irreversibility of M-phase transitions by controlling reversible phosphorylations, feedback loops and thresholds, primarily governed by Cdk1 and its counteracting phosphatases. At the center of the impressive number of phosphorylations occurring upon M-phase entry (over 32,000), a small protein, ARPP19, orientates the cellular decision to divide or not according to the phosphorylation of two of its residues. Determining the molecular targets of the S109-phosphorylated forms of ARPP19, as well as identifying the phosphatase that regulates the S109-phosphorylation level of ARPP19 and the kinase that collaborates with Gwl to phosphorylate S109 of ARPP19 will enlighten our understanding not only of the M-phase control but also of the mechanisms of cAMP-regulated signal transduction pathways.

Acknowledgements

We apologize to those whose work is omitted owing to space constraints. This work was supported by grants from CNRS, University Pierre and Marie Curie, and ANR (ANR-13-BSV2-0008-01 to CJ).

Author details

Aude Dupré and Catherine Jessus*

*Address all correspondence to: catherine.jessus@upmc.fr

Sorbonne University, UPMC Univ Paris 06, CNRS, Laboratory of Developmental Biology - Institute Biology Paris Seine (LBD - IBPS), Paris, France

References

[1] Masui Y, Markert CL. Cytoplasmic control of nuclear behavior during meiotic maturation of frog oocytes. The Journal of Experimental Zoology. Jun, 1971;**177**:129

[2] Wasserman WJ, Masui Y. Effects of cycloheximide on a cytoplasmic factor initiating meiotic maturation in *Xenopus* oocytes. Experimental Cell Research. 1975;**91**:381

[3] Kishimoto T. Entry into mitosis: A solution to the decades-long enigma of MPF. Chromosoma. Dec, 2015;**124**:417

[4] Reynhout JK, Smith LD. Studies on the appearance and nature of a maturation-inducing factor in the cytoplasm of amphibian oocytes exposed to progesterone. Developmental Biology. 1974;**38**:394

[5] Kishimoto T, Usui N, Kanatani H. Breakdown of starfish ovarian follicle induced by maturation-promoting factor. Developmental Biology. Jan, 1984;**101**:28

[6] Kishimoto T, Yamazaki K, Kato Y, Koide S, Kanatani H. Induction of starfish oocyte maturation by maturation-promoting factor of mouse and surf clam oocytes. The Journal of Experimental Zoology. 1984;**231**:293

[7] Hartwell LH, Culotti J, Reid B. Genetic control of the cell-division cycle in yeast. I. Detection of mutants. Proceedings of the National Academy of Sciences of the United States of America. 1970;**66**:352

[8] Nurse P, Thuriaux P, Nasmyth K. Genetic control of the cell division cycle in the fission yeast *Schizosaccharomyces pombe*. Molecular & General Genetics. 1976;**146**:167

[9] Izawa D, Pines J. How APC/C-Cdc20 changes its substrate specificity in mitosis. Nature Cell Biology. Mar, 2011;**13**:223

[10] Lindon C. Control of mitotic exit and cytokinesis by the APC/C. Biochemical Society Transactions. Jun, 2008;**36**:405

[11] Virshup DM, Shenolikar S. From promiscuity to precision: Protein phosphatases get a makeover. Molecular Cell. Mar 13, 2009;**33**:537

[12] Kim HS, Fernandes G, Lee CW. Protein phosphatases involved in regulating mitosis: Facts and hypotheses. Molecules and Cells. Sep, 2016;**39**:654

[13] Queralt E, Uhlmann F. Cdk-counteracting phosphatases unlock mitotic exit. Current Opinion in Cell Biology. Dec, 2008;**20**:661

[14] Bollen M, Gerlich DW, Lesage B. Mitotic phosphatases: From entry guards to exit guides. Trends in Cell Biology. Oct, 2009;**19**:531

[15] Heim A, Rymarczyk B, Mayer TU. Regulation of cell division. Advances in Experimental Medicine and Biology. 2017;**953**:83

[16] Gowdy PM, Anderson HJ, Roberge M. Entry into mitosis without Cdc2 kinase activation. Journal of Cell Science. Nov, 1998;**111(Pt 22)**:3401

[17] Goris J, Hermann J, Hendrix P, Ozon R, Merlevede W. Okadaic acid, a specific protein phosphatase inhibitor, induces maturation and MPF formation in *Xenopus laevis* oocytes. FEBS Letters. 1989;**245**:91

[18] Skoufias DA, Indorato RL, Lacroix F, Panopoulos A, Margolis RL. Mitosis persists in the absence of Cdk1 activity when proteolysis or protein phosphatase activity is suppressed. The Journal of Cell Biology. Nov 19, 2007;**179**:671

[19] Mochida S, Ikeo S, Gannon J, Hunt T. Regulated activity of PP2A-B55delta is crucial for controlling entry into and exit from mitosis in Xenopus egg extracts. The EMBO Journal. Aug 20, 2009;**28**:2777

[20] Gharbi-Ayachi A, et al. The substrate of Greatwall kinase, Arpp19, controls mitosis by inhibiting protein phosphatase 2A. Science. Dec 17, 2010;**330**:1673

[21] Mochida S, Maslen SL, Skehel M, Hunt T. Greatwall phosphorylates an inhibitor of protein phosphatase 2A that is essential for mitosis. Science. Dec 17, 2010;**330**:1670

[22] Vigneron S, et al. Greatwall maintains mitosis through regulation of PP2A. EMBO Journal. Aug 13, 2009;**28**:2786

[23] White-Cooper H, Carmena M, Gonzalez C, Glover DM. Mutations in new cell cycle genes that fail to complement a multiply mutant third chromosome of drosophila. Genetics. Nov, 1996;**144**:1097

[24] Yu J, et al. Greatwall kinase: a nuclear protein required for proper chromosome condensation and mitotic progression in Drosophila. Journal of Cell Biology. Feb 16, 2004;**164**:487

[25] Yu J, Zhao Y, Li Z, Galas S, Goldberg ML. Greatwall kinase participates in the CDC2 autoregulatory loop in Xenopus egg extracts. Molecular Cell. Apr 7, 2006;**22**:83

[26] Vigneron S, et al. Characterization of the mechanisms controlling Greatwall activity. Molecular and Cellular Biology. Jun, 2011;**31**:2262

[27] Blake-Hodek KA, et al. Determinants for activation of the atypical AGC kinase Greatwall during M phase entry. Molecular and Cellular Biology. Apr, 2012;**32**:1337

[28] Vigneron S, et al. The master Greatwall kinase, a critical regulator of mitosis and meiosis. The International Journal of Developmental Biology. 2016;**60**:245

[29] Bontron S, et al. Yeast endosulfines control entry into quiescence and chronological life span by inhibiting protein phosphatase 2A. Cell Reports. Jan 31, 2013;**3**:16

[30] Talarek N, Bontron S, De Virgilio C. Quantification of mRNA stability of stress-responsive yeast genes following conditional excision of open reading frames. RNA Biology. Aug, 2013;**10**:1299

[31] Talarek N, et al. Initiation of the TORC1-regulated G0 program requires Igo1/2, which license specific mRNAs to evade degradation via the 5′-3′ mRNA decay pathway. Molecular Cell. May 14, 2010;**38**:345

[32] Juanes MA, et al. Budding yeast greatwall and endosulfines control activity and spatial regulation of PP2A(Cdc55) for timely mitotic progression. PLoS Genetics. 2013;**9**:e1003575

[33] White RE, Giffard RG. MicroRNA-320 induces neurite outgrowth by targeting ARPP-19. Neuroreport. Jul 11, 2012;**23**:590

[34] Jiang T, Zhao B, Li X, Wan J. ARPP-19 promotes proliferation and metastasis of human glioma. Neuroreport. Sep 07, 2016;**27**:960

[35] Williams BC, et al. Greatwall-phosphorylated Endosulfine is both an inhibitor and a substrate of PP2A-B55 heterotrimers. eLife. 2014;**3**:e01695

[36] Luo X, Talarek N, De Virgilio C. Initiation of the yeast G0 program requires Igo1 and Igo2, which antagonize activation of decapping of specific nutrient-regulated mRNAs. RNA Biology. Jan-Feb, 2011;**8**:14

[37] Wong PY, Ma HT, Lee HJ, Poon RY. MASTL(Greatwall) regulates DNA damage responses by coordinating mitotic entry after checkpoint recovery and APC/C activation. Scientific Reports. 2016;**6**:22230

[38] Okumura E, et al. Cyclin B-Cdk1 inhibits protein phosphatase PP2A-B55 via a Greatwall kinase-independent mechanism. The Journal of Cell Biology. Mar 17, 2014;**204**:881

[39] Burgess A, et al. Loss of human Greatwall results in G2 arrest and multiple mitotic defects due to deregulation of the cyclin B-Cdc2/PP2A balance. Proceedings of the National Academy of Sciences of the United States of America. Jul 13, 2010;**107**:12564

[40] Voets E, Wolthuis RM. MASTL is the human orthologue of Greatwall kinase that facilitates mitotic entry, anaphase and cytokinesis. Cell Cycle. Sep 29, 2010;**9**:3591

[41] Alvarez-Fernandez M, et al. Greatwall is essential to prevent mitotic collapse after nuclear envelope breakdown in mammals. Proceedings of the National Academy of Sciences of the United States of America. Oct 22, 2013;**110**:17374

[42] Cundell MJ, et al. The BEG (PP2A-B55/ENSA/Greatwall) pathway ensures cytokinesis follows chromosome separation. Molecular Cell. Nov 7, 2013;**52**:393

[43] Manchado E, et al. Targeting mitotic exit leads to tumor regression in vivo: Modulation by Cdk1, Mastl, and the PP2A/B55alpha,delta phosphatase. Cancer Cell. Dec 14, 2010;**18**:641

[44] Heim A, Konietzny A, Mayer TU. Protein phosphatase 1 is essential for Greatwall inactivation at mitotic exit. EMBO Reports. Nov, 2015;**16**:1501

[45] Ma S, et al. Greatwall dephosphorylation and inactivation upon mitotic exit is triggered by PP1. Journal of Cell Science. Apr 1, 2016;**129**:1329

[46] Wu JQ, et al. PP1-mediated dephosphorylation of phosphoproteins at mitotic exit is controlled by inhibitor-1 and PP1 phosphorylation. Nature Cell Biology. May, 2009;**11**:644

[47] Russell P, Nurse P. cdc25+ functions as an inducer in the mitotic control of fission yeast. Cell. 1986;**45**:145

[48] Simanis V, Nurse P. The cell cycle control gene cdc2+ of fission yeast encodes a protein kinase potentially regulated by phosphorylation. Cell. 1986;**45**:261

[49] Mueller PR, Coleman TR, Kumagai A, Dunphy WG. Myt1: A membrane-associated inhibitory kinase tht phosphorylates cdc2 on both threonine-14 and tyrosine-15. Science. 1995;**270**:86

[50] Booher RN, Holman PS, Fattaey A. Human Myt1 is a cell cycle-regulated kinase that inhibits Cdc2 but not Cdk2 activity. The Journal of Biological Chemistry. 1997;**272**:22300

[51] Fattaey A, Booher RN. Myt1: A Wee1-type kinase that phosphorylates Cdc2 on residue Thr14. Progress in Cell Cycle Research. 1997;**3**:233

[52] Liu F, Stanton JJ, Wu ZQ, PiwnicaWorms H. The human Myt1 kinase preferentially phosphorylates Cdc2 on threonine 14 and localizes to the endoplasmic reticulum and Golgi complex. Molecular and Cellular Biology. 1997;**17**:571

[53] Malumbres M. Cyclin-dependent kinases. Genome Biology. 2014;**15**:122

[54] Hegarat N, Rata S, Hochegger H. Bistability of mitotic entry and exit switches during open mitosis in mammalian cells. BioEssays. Jul, 2016;**38**:627

[55] Lorca T, et al. Constant regulation of both the MPF amplification loop and the Greatwall-PP2A pathway is required for metaphase II arrest and correct entry into the first embryonic cell cycle. Journal of Cell Science. Jul 1, 2010;**123**:2281

[56] Schmitz MH, et al. Live-cell imaging RNAi screen identifies PP2A-B55alpha and importin-beta1 as key mitotic exit regulators in human cells. Nature Cell Biology. Aug 15, 2010;**12**:886

[57] Blackford AN, Jackson SP. ATM, ATR, and DNA-PK: The trinity at the heart of the DNA damage response. Molecular Cell. Jun 15, 2017;**66**:801

[58] Taylor SS, Ilouz R, Zhang P, Kornev AP. Assembly of allosteric macromolecular switches: Lessons from PKA. Nature Reviews. Molecular Cell Biology. Oct, 2012;**13**:646

[59] Dema A, Perets E, Schulz MS, Deak VA, Klussmann E. Pharmacological targeting of AKAP-directed compartmentalized cAMP signalling. Cellular Signalling. Dec, 2015; **27**:2474

[60] Sheppard JR, Prescott DM. Cyclic AMP levels in synchronized mammalian cells. Experimental Cell Research. Nov, 1972;**75**:293

[61] Kurokawa K, Kato J. Cyclic AMP delays G2 progression and prevents efficient accumulation of cyclin B1 proteins in mouse macrophage cells. Cell Structure and Function. Dec, 1998;**23**:357

[62] Dicitore A, et al. The cAMP analogs have potent anti-proliferative effects on medullary thyroid cancer cell lines. Endocrine. Jan, 2016;**51**:101

[63] Rodriguez-Collazo P, et al. cAMP signaling regulates histone H3 phosphorylation and mitotic entry through a disruption of G2 progression. Experimental Cell Research. Sep 10, 2008;**314**:2855

[64] Dupre A, Daldello EM, Nairn AC, Jessus C, Haccard O. Phosphorylation of ARPP19 by protein kinase A prevents meiosis resumption in Xenopus oocytes. Nature Communications. 2014;**5**:3318

[65] Smith LD, Ecker RE, Subtelny S. In vitro induction of physiological maturation in *Rana pipiens* oocytes removed from their ovarian follicles. Developmental Biology. Jun, 1968;**17**:627

[66] Masui Y. Relative roles of the pituitary, follicle cells, and progesterone in the induction of oocyte maturation in Rana Pipiens. The Journal of Experimental Zoology. 1967;**166**:365

[67] Schuetz A. Effect of steroids on germinal vesicle of oocytes of the frog (*Rana pipiens*) in vitro. Proceedings of the Society for Experimental Biology and Medicine. 1967;**124**:1307

[68] Fortune JE, Concannon PW, Hansel W. Ovarian progesterone levels during in vitro oocyte maturation and ovulation in *Xenopus laevis*. Biology of Reproduction. 1975;**13**:561

[69] Haccard O, et al. Naturally occurring steroids in Xenopus oocyte during meiotic maturation. Unexpected presence and role of steroid sulfates. Molecular and Cellular Endocrinology. Oct 15, 2012;**362**:110

[70] Karaiskou A, Dupre A, Haccard O, Jessus C. From progesterone to active Cdc2 in *Xenopus* oocytes: A puzzling signalling pathway. Biology of the Cell. Sep, 2001;**93**:35

[71] Inoue D, Ohe M, Kanemori Y, Nobui T, Sagata N. A direct link of the Mos-MAPK pathway to Erp1/Emi2 in meiotic arrest of *Xenopus laevis* eggs. Nature. Apr 26, 2007;**446**:1100

[72] Rauh NR, Schmidt A, Bormann J, Nigg EA, Mayer TU. Calcium triggers exit from meiosis II by targeting the APC/C inhibitor XErp1 for degradation. Nature. Oct 13, 2005;**437**:1048

[73] Nishiyama T, Ohsumi K, Kishimoto T. Phosphorylation of Erp1 by p90rsk is required for cytostatic factor arrest in *Xenopus laevis* eggs. Nature. Apr 26, 2007;**446**:1096

[74] Liu J, Maller JL. Calcium elevation at fertilization coordinates phosphorylation of XErp1/Emi2 by Plx1 and CaMK II to release metaphase arrest by cytostatic factor. Current Biology. Aug 23, 2005;**15**:1458

[75] Kobayashi H, et al. On the synthesis and destruction of A- and B-type cyclins during oogenesis and meiotic maturation in *Xenopus laevis*. The Journal of Cell Biology. Aug, 1991;**114**:755

[76] De Smedt V, et al. Thr-161 phosphorylation of monomeric Cdc2. Regulation by protein phosphatase 2C in Xenopus oocyte. Journal of Biological Chemistry. Aug 2, 2002;**277**:28592

[77] Gaffre M, et al. A critical balance between Cyclin B synthesis and Myt1 activity controls meiosis entry in Xenopus oocytes. Development. Sep, 2011;**138**:3735

[78] Ruiz EJ, Hunt T, Nebreda AR. Meiotic inactivation of Xenopus Myt1 by CDK/XRINGO, but not CDK/cyclin, via site-specific phosphorylation. Molecular Cell. Oct 24, 2008;**32**:210

[79] Okumura E, et al. Akt inhibits Myt1 in the signalling pathway that leads to meiotic G2/M-phase transition. Nature Cell Biology. Feb, 2002;**4**:111

[80] Duckworth BC, Weaver JS, Ruderman JV. G2 arrest in *Xenopus* oocytes depends on phosphorylation of cdc25 by protein kinase A. Proceedings of the National Academy of Sciences of the United States of America. Dec 24, 2002;**99**:16794

[81] Mulner O, Cartaud A, Ozon R. Cyclic AMP phosphodiesterase activities in *Xenopus laevis* oocytes. Differentiation. 1980;**16**:31

[82] Mulner O, Huchon D, Thibier C, Ozon R. Cyclic AMP synthesis in *Xenopus laevis* oocytes: Inhibition by progesterone. Biochimica et Biophysica Acta. 1979;**582**:179

[83] Sadler SE, Maller JL. Progesterone inhibits adenylate cyclase in *Xenopus* oocytes. Action on the guanine nucleotide regulatory protein. The Journal of Biological Chemistry. 1981;**256**:6368

[84] Sadler SE, Maller JL. Inhibition of *Xenopus* oocyte adenylate cyclase by progesterone: A novel mechanism of action. Advances in Cyclic Nucleotide and Protein Phosphorylation Research. 1985;**19**:179

[85] Sadler SE, Maller JL. Vivo regulation of cyclic AMP phosphodiesterase in *Xenopus* oocytes. Stimulation by insulin and insulin-like growth factor 1. The Journal of Biological Chemistry. 1987;**262**:10644

[86] Finidori J, Hanoune J, Baulieu EE. Adenylate cyclase in *Xenopus laevis* oocytes: Characterization of the progesterone-sensitive, membrane-bound form. Molecular and Cellular Endocrinology. 1982;**28**:211

[87] Finidori-lepicard J, Schorderet-slatkine S, Hanoune J, Baulieu EE. Progesterone inhibits membrane bound adenylate cyclase in *Xenopus laevis* oocytes. Nature. 1981;**292**:255

[88] Schorderet-slatkine S, Baulieu E-E. Forskolin increases cAMP and inhibits progesterone induced meiosis reinitiation in *Xenopus laevis* oocytes. Endocrinology. 1982;**111**:1385

[89] Schorderet-Slatkine S, Schorderet M, Baulieu EE. Cyclic AMP-mediated control of meiosis: Effects of progesterone, cholera toxin, and membrane-active drugs in *Xenopus laevis* oocytes. Proceedings of the National Academy of Sciences of the United States of America. 1982;**79**:850

[90] Maller JL, Butcher FR, Krebs EG. Early effect of progesterone on levels of cyclic adenosine 3′:5′-monophosphate in *Xenopus* oocytes. The Journal of Biological Chemistry. 1979;**254**:579

[91] Rime H, Haccard O, Ozon R. Activation of p34cdc2 kinase by cyclin is negatively regulated by cyclic amp-dependent protein kinase in Xenopus oocytes. Developmental Biology. May, 1992;**151**:105

[92] Maller JL, Krebs EG. Progesterone-stimulated meiotic cell division in *Xenopus* oocytes. Induction by regulatory subunit and inhibition by catalytic subunit of adenosine 3′:5′-monophosphate-dependent protein kinase. The Journal of Biological Chemistry. 1977;**252**:1712

[93] Huchon D, Ozon R, Fischer EH, Demaille JG. The pure inhibitor of cAMP-dependent protein kinase initiates *Xenopus laevis* meiotic maturation. A 4-step scheme for meiotic maturation. Molecular and Cellular Endocrinology. May, 1981;**22**:211

[94] Dupre A, Haccard O, Jessus C. Mos in the oocyte: How to use MAPK independently of growth factors and transcription to control meiotic divisions. Journal of Signal Transduction. 2011;**2011**:350412

[95] Amiel A, Leclere L, Robert L, Chevalier S, Houliston E. Conserved functions for Mos in eumetazoan oocyte maturation revealed by studies in a cnidarian. Current Biology. Feb 24, 2009;**19**:305

[96] Dumollard R, et al. Mos limits the number of meiotic divisions in urochordate eggs. Development. Mar, 2011;**138**:885

[97] Haccard O, Jessus C. Oocyte maturation, Mos and cyclins—A matter of synthesis: Two functionally redundant ways to induce meiotic maturation. Cell Cycle. Jun, 2006;**5**:1152

[98] Haccard O, Jessus C. Redundant pathways for Cdc2 activation in *Xenopus* oocyte: Either cyclin B or Mos synthesis. EMBO Reports. Mar, 2006;**7**:321

[99] Frank-Vaillant M, Jessus C, Ozon R, Maller JL, Haccard O. Two distinct mechanisms control the accumulation of cyclin B1 and mos in *Xenopus* oocytes in response to progesterone. Molecular Biology of the Cell. 1999;**10**:3279

[100] Castro A, et al. Cyclin B/cdc2 induces c-Mos stability by direct phosphorylation in Xenopus oocytes. Molecular Biology of the Cell. Sep, 2001;**12**:2660

[101] Nebreda AR, Gannon JV, Hunt T. Newly synthesized protein(s) must associate with p34cdc2 to activate MAP kinase and MPF during progesterone-induced maturation of *Xenopus* oocytes. The EMBO Journal. 1995;**14**:5597

[102] Wang R, et al. Regulation of Cdc25C by ERK-MAP kinases during the G(2)/M transition. Cell. Mar 23, 2007;**128**:1119

[103] Matten W, Daar I, Vande Woude GF. Protein kinase A acts at multiple points to inhibit *Xenopus* oocyte maturation. Molecular and Cellular Biology. 1994;**14**:4419

[104] Rime H, et al. Characterization of MPF activation by okadaic acid in *Xenopus* oocyte. Cell Differentiation and Development. 1990;**29**:47

[105] Yamamoto TM, et al. Regulation of Greatwall kinase during Xenopus oocyte maturation. Molecular Biology of the Cell. Jul 1, 2011;**22**:2157

[106] Hara M, et al. Greatwall kinase and cyclin B-Cdk1 are both critical constituents of M-phase-promoting factor. Nature Communications. 2012;**3**:1059

[107] Wang J, Cao WL, Liu XJ. Protein kinase a(PKA)-restrictive and PKA-permissive phases of oocyte maturation. Cell Cycle. Jan, 2006;**5**:213

[108] Wang J, Liu XJ. Progesterone inhibits protein kinase A (PKA) in *Xenopus* oocytes: Demonstration of endogenous PKA activities using an expressed substrate. Journal of Cell Science. Oct 1, 2004;**117**:5107

[109] Zhao Y, et al. Roles of greatwall kinase in the regulation of cdc25 phosphatase. Molecular Biology of the Cell. Apr, 2008;**19**:1317

[110] Dupre A, et al. The phosphorylation of ARPP19 by Greatwall renders the auto-amplification of MPF independently of PKA in Xenopus oocytes. Journal of Cell Science. Sep 1, 2013;**126**:3916

[111] Schmitt A, Nebreda AR. Inhibition of *Xenopus* oocyte meiotic maturation by catalytically inactive protein kinase A. Proceedings of the National Academy of Sciences of the United States of America. Apr 2, 2002;**99**:4361

[112] Han SJ, Chen R, Paronetto MP, Conti M. Wee1B is an oocyte-specific kinase involved in the control of meiotic arrest in the mouse. Current Biology. Sep 20, 2005;**15**:1670

[113] Han SJ, Conti M. New pathways from PKA to the Cdc2/cyclin B complex in oocytes: Wee1B as a potential PKA substrate. Cell Cycle. Feb, 2006;**5**:227

[114] Nakajo N, et al. Absence of Wee1 ensures the meiotic cell cycle in Xenopus oocytes. Genes & Development. 2000;**14**:328

[115] Boyer J, Asselin J, Belle R, Ozon R. Progesterone and cAMP-dependent protein kinase regulate in vivo the level of phosphorylation of two proteins (Mr 20,000 and Mr 32,000) in Xenopus oocytes. Developmental Biology. Feb, 1986;**113**:420

[116] Mochida S. Regulation of alpha-endosulfine, an inhibitor of protein phosphatase 2A, by multisite phosphorylation. The FEBS Journal. Feb, 2014;**281**:1159

[117] Dulubova I, et al. ARPP-16/ARPP-19: A highly conserved family of cAMP-regulated phosphoproteins. Journal of Neurochemistry. Apr, 2001;**77**:229

[118] Girault JA. Integrating neurotransmission in striatal medium spiny neurons. Advances in Experimental Medicine and Biology. 2012;**970**:407

[119] Girault JA. Signaling in striatal neurons: The phosphoproteins of reward, addiction, and dyskinesia. Progress in Molecular Biology and Translational Science. 2012;**106**:33

[120] Ysselstein D, et al. Endosulfine-alpha inhibits membrane-induced alpha-synuclein aggregation and protects against alpha-synuclein neurotoxicity. Acta Neuropathologica Communications. Jan 10, 2017;**5**:3

[121] Dupre AI, Haccard O, Jessus C. The greatwall kinase is dominant over PKA in controlling the antagonistic function of ARPP19 in Xenopus oocytes. Cell Cycle. 2017;**16**:1440

[122] Andrade EC, et al. ARPP-16 is a striatal-enriched inhibitor of protein phosphatase 2A regulated by microtubule-associated serine/threonine kinase 3 (Mast 3 kinase). The Journal of Neuroscience. Feb 06, 2017;**37**:2709

[123] Musante V, et al. Reciprocal regulation of ARPP-16 by PKA and MAST3 kinases provides a cAMP-regulated switch in protein phosphatase 2A inhibition. eLife. Jun 14, 2017;**6**:e24998

[124] Huchon D, Ozon R, Demaille JG. Protein phosphatase-1 is involved in *Xenopus* oocyte maturation. Nature. 1981;**294**:358

Chapter 2

The Mitotic Protein Kinase Haspin and Its Inhibitors

Omid Feizbakhsh, Matthieu Place, Xavier Fant,
Frédéric Buron, Sylvain Routier and
Sandrine Ruchaud

Additional information is available at the end of the chapter

http://dx.doi.org/10.5772/intechopen.70732

Abstract

Haspin is an atypical serine/threonine protein kinase essential to mitosis. Unlike other protein kinases, its kinase domain does not require phosphorylation in order to be activated and bears very high substrate specificity and selectivity. Few substrates have been identified so far. Haspin phosphorylation on threonine 3 of Histone H3 from prophase to anaphase participates to centromeric Aurora B localization and ensures proper kinetochore-microtubule attachment. Haspin is also involved in the maintenance of centromeric cohesion and the mitotic spindle. Inhibitors have been developed and provided tools to dissect Haspin function. The kinase is now considered as a potential therapeutic target against cancer. We discuss here the latest findings on this essential mitotic protein.

Keywords: Haspin kinase, mitosis, inhibitors, H3T3ph, centromere

1. Introduction

Protein kinases play an important role in cell cycle regulation. Together with protein phosphatases, they regulate the phosphorylation status of thousands of substrates, including proteins ensuring cell cycle progression. Cell division, that is, mitosis, is a crucial step of the cell cycle and is essential to genomic stability. Mistakes during this process can cause various developmental diseases and cancers. Its orchestration is highly regulated by various families of protein kinases, including cyclin-dependent kinases (Cdk), Aurora kinases, polo-like kinases (Plks), and NimA-related kinases (Neks) whose roles in mitosis are well documented [1, 2]. Haspin is a serine/threonine kinase discovered in the early 1990s essential to mitosis. Despite recent progress, regulation of its activity and its biological functions is still poorly understood. Haspin is involved in chromosome alignment, centromeric cohesion, and spindle stability making it a potential target against cancer. Latest data from the literature concerning this protein kinase as well as pharmacological inhibitors are presented.

2. Discovery

Haspin mRNA was first detected in mouse germ cells in 1994 and named germ cell–specific gene 2 (*GSG2*) [3]. The coded protein localizes in the nucleus of germ cells and showed kinase activity. It was subsequently renamed haploid germ cell–specific nuclear protein kinase (Haspin) [4]. *Haspin* mRNA is found in diploid cells of many proliferative tissues such as testis, thymus, bone marrow, and spleen as well as in many proliferative cell lines [5–7]. Its expression is comparatively reduced in somatic tissues. In addition, Haspin orthologs were found in several eukaryotes, such as yeasts, plants, flies, fishes and mammals, and a large group in *Caenorhabditis elegans*. Phylogenetic analysis indicates that Haspin proteins form a new family of eukaryotic proteins kinases (ePK) [5].

3. Haspin, an atypical kinase structure

Human Haspin is a 798 amino acid serine/threonine protein kinase. The N-terminal part (aa 1–469) is the less conserved among species and thought to act as a regulatory domain. The well-conserved C-terminal part (aa 470–798) corresponds to the catalytic kinase domain [8–10]. To date, Haspin kinase domain was crystallized in the presence of ATP analogs such as 5-iodotubercidin (5-ITu) or in the presence of a specific substrate of the kinase, Histone H3 [8, 10, 11]. Disorganization of the N-terminal domain prevented, so far, crystallization of the entire protein.

The structure of human Haspin kinase domain (aa 470–798) shows similarity to the kinase domain from other kinases of the ePK family. As most protein kinases, it includes a small lobe on the N-terminal side and a large lobe on the C-terminal side. A substrate-binding site and an ATP binding pocket are found between the two lobes. The catalytic domain of Haspin displays specific structural features that are not observed in other members of the ePK family.

Compared to canonical protein kinases, Haspin structure revealed several unique and specific structural features that are highly conserved in several species. These characteristics result from amino acid insertion, deletion, or changes in the protein sequence of its catalytic domain in comparison to other ePKs [8, 10]. The structure of most protein kinases is generally very dynamic, allowing a kinase to transform, by a conformational change, from an inactive to an active state through phosphorylation or interaction with a partner [12]. On the contrary, Haspin kinase domain is rigid and fixed in a constitutively active conformational state [8, 10, 11] (**Figure 1A**).

This stability is achieved by the combination of different structural elements. Mainly, the generally mobile glycine rich P-loop is stabilized in Haspin by the insertion of an additional helix in the upper lobe called either upper lobe helix (ulH) [8] or $\alpha C'$ helix [10]. This helix insertion is mostly conserved throughout Haspin orthologs apart from the fission yeast Alk1 and Alk2. Usually, mobile αC helix in the small N-terminal lobe is also stabilized by a number of hydrophobic contacts. Haspin activation segment is another atypical structural element. In most kinases, the activation segment has a regulatory purpose, acquiring an active

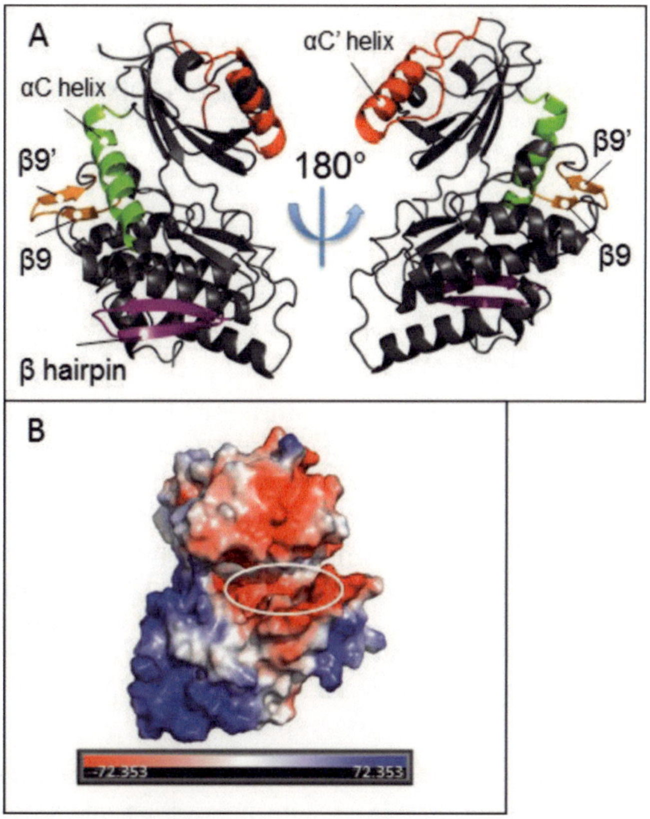

Figure 1. Haspin kinase domain 3D structure. (A) Insertion elements, αC′ helix, β9 and β9′, and β hairpin are indicated. (B) Representation of Haspin surface electrostatic charges; the negatively charged substrate-binding site is circled.

conformation upon phosphorylation and allowing substrate binding. Haspin activation segment is stabilized in a constitutively active conformation [8, 10].

Finally, Haspin bears a very specific substrate-binding site. Knowing that histone H3 tail (positively charged) is a specific Haspin substrate suggested a negatively charged binding site as depicted on **Figure 1B**. Maiolica *et al.* provided insights into this peculiar substrate-binding site resolving the crystal structure of Haspin kinase domain bond to the first seven residues of Histone H3 [11]. The study revealed that three residues of the latter, Ala1, Arg2, and the phospo-acceptor site Thr3, are deeply anchored in the substrate hydrophilic binding site of Haspin. These peculiarities create a highly selective substrate-binding site [11].

4. Haspin substrates

Very few substrates of Haspin have been identified and characterized so far. Histone H3 was the first Haspin substrate to be identified. It is specifically phosphorylated on Thr3 [8, 10, 13].

This phosphorylation (H3T3ph) was demonstrated both in vitro and in several cell lines by immunofluorescence, using histone H3 Thr3 phospho-specific antibodies (**Figure 5** upper panel). Haspin depletion by siRNA eliminates H3T3 phosphorylation in mitotic cells, and ectopic overexpression of Haspin leads to abnormal H3T3 phosphorylation levels in interphase cells confirming that H3T3ph is specific of Haspin activity [7, 14, 15]. H3 is phosphorylated on Thr3 through most of mitosis. Kurihara *et al.* showed that *Arabidopsis thaliana* Haspin, AtHAspin, is an H3T3ph kinase. They have further demonstrated that AtHaspin phosphorylates both Thr3 and Thr11 of Histone H3 *in vitro* [16, 17]. It is to be noted that Haspin-homologous proteins in budding yeast, Alk1 and Alk2, have not shown any ability to phosphorylate histones [18], whereas the fission yeast Haspin-related kinase, Hrk1, has been shown to be the major H3T3 phosphorylating kinase in this species [19].

Histone macroH2A is an histone variant found enriched on inactive X chromosome of female mammals [20]. Several studies demonstrated that histone macroH2A functions both as a positive and a negative regulator of gene transcription. A phosphoproteomic study showed that inhibition of Haspin by 5-ITu led to a sharp decrease in serine phosphorylation of histone macroH2A [11]. This phosphorylation was confirmed in vitro and in HEK293 cells, where overexpression of Haspin caused hyper-phosphorylation of Histone macroH2A on Ser137, the latter being inhibited by 5-ITu. It has also been reported that the macro domain of Histone macroH2A controls the levels of Ser10 and Thr3 phosphorylation of histone H3 in human cells and would be involved in controlling chromatin condensation [21, 22]. The functionality of Haspin in these mechanisms remains to be confirmed.

CENP-T is a component of the constitutive centromere-associated network (CCAN), which plays a central role in kinetochore assembly, mitotic progression, and segregation of chromosomes [23]. CENP-T has been identified as substrate of Haspin by consensus site prediction (see below), and its phosphorylation on several sites confirmed by in vitro kinase assay [11].

The Haspin kinase substrate recognition motif has been determined by positional scanning-oriented peptide library screening (PS-OPLS) as A/V-R-T/S-K-(X-no D/E) with a preference for threonine residues [11]. Acidic residues have been shown to impair Haspin recognition when in the surrounding of the phosphorylation site [11].

5. Haspin biological function

5.1. Haspin localization

Haspin is constitutively expressed throughout the cell cycle, unlike other mitotic kinases such as Aurora B and Plk1, which are degraded at the end of mitosis [13, 24, 25]. So far, the precise cellular localization of the endogenous protein could not be determined due to lack of immunofluorescence-specific antibodies to Haspin. However, several overexpression studies have reported localization of GFP- or Myc-tagged Haspin in different eukaryotic cell lines (HeLa, U-2 OS, Hek293, COS-7) using time-lapse video microscopy or immunofluorescence staining techniques or time-lapse video microscopy [4, 13]. All these studies showed that Haspin localizes in discrete foci and nucleoli in the nucleus during interphase. It is to be noted

http://dx.doi.org/10.5772/intechopen.70732

that the N-terminal domain of human Haspin exhibits two potential nuclear localization signals (NLSs) that are conserved in mice and rats [9]. Localization pattern in mitosis is more complex (**Figure 2**).

At the end of G2/onset of prophase, Haspin appears associated with condensed chromosomes until anaphase B. Myc:Haspin is observed along chromosome arms with a clear concentration at centromeres [13]. GFP:Haspin was also detected at the centrosomes and mitotic spindle in prometaphase cells until telophase, where a weak signal is detected in the midbody [13]. Nuclear and chromosomal localization throughout mitosis of endogenous Haspin:YFP knocked-in have been recently confirmed by video microscopy [26]. **Figure 2** shows the location of Haspin during the various stages of the cell cycle with emphasis on mitosis (**Figure 2**). Phosphorylation of histone H3 on Thr3 was also examined in plants. In most of the species studied, it appears on chromosomes at the end of G2 phase and disappears during anaphase [27], in contrast to mammals, phosphorylation of Thr3 is seen primarily at pericentromeres in prophase and then along chromosome arms during prometaphase [27, 28].

5.2. Specific localization of Haspin during meiosis

GFP:Haspin co-localizes with chromatin and H3T3ph during all meiotic stages [29, 30]. Notably, H3T3ph levels were increased with overexpression of exogenous Haspin, thus confirming that Haspin phosphosphorylates H3T3 in oocytes. During metaphase I, Haspin is detected at the centromeres and along sister chromatids. After metaphase I, GFP:Haspin is also located in a discrete region of the oocyte's cortex which is in the immediate vicinity of chromatin and the spindle [29]. Haspin was also weakly detected as filamentous aggregates on the spindle.

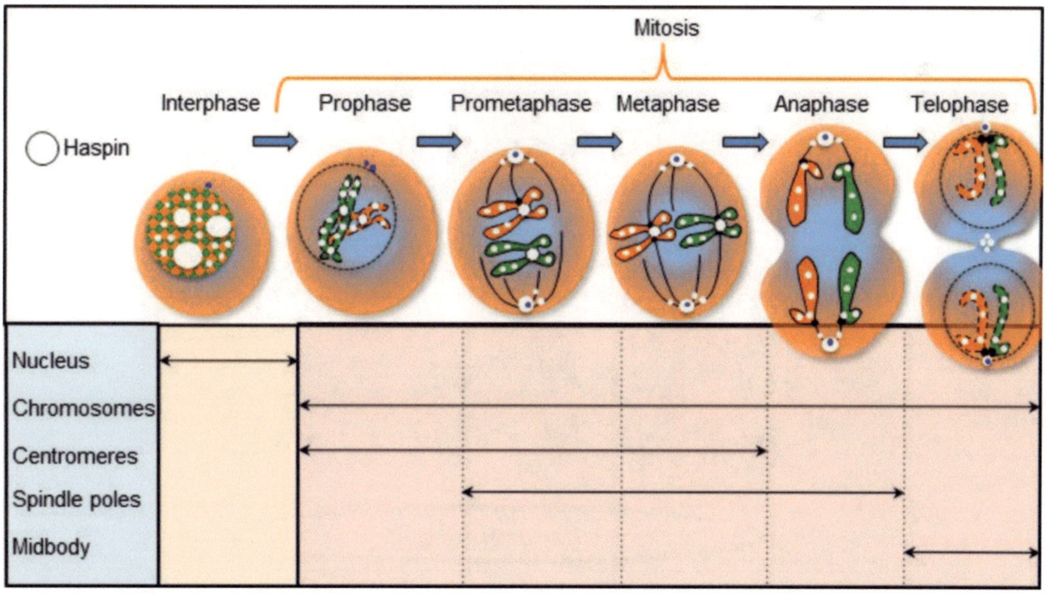

Figure 2. Haspin reported localization along cell cycle in mammal cells.

At anaphase I/telophase I transition, Haspin is translocated from chromosomes to midbody [29, 30].

5.3. How is Haspin recruited onto chromosomes?

A yeast two-hybrid screen of cohesin-related proteins on the fission yeast Haspin homolog Hrk1 identified an interaction with the cohesin-associated protein Pds5 [19]. The interaction was confirmed in cells using a model in which Pds5:mCherry:LacI was tethered onto a specific location on a chromosome arm through a LacO/LacI system onto which Hrk1:GFP was shown to co-localize (**Figure 3**) [19].

Vertebrates have two version of Pds5 protein, Pds5A and B [31, 32]. Carretero *et al.* demonstrated that Pds5B-deficient MEF cells showed a decreased activation of Aurora B and Haspin at centromeres and an impaired centromeric localization of Aurora B suggesting that Pds5B may be involved in the recruitment of Haspin on centromeres [33]. The recruitment of Haspin by Pds5B has been confirmed in human cell lines by Hindriksen *et al.* [26]. Using a LacO/LacI system in which Pds5B:RFP:LacI was shown to recruit Haspin:YFP on a LacO repeats inserted on chromosome 1 of human U-2 OS cells [26].

Two recent studies in Xenopus and yeast showed that SUMOylated DNA topoisomerase IIα C-terminal domain can bind Haspin and regulate its localization at centromeres [34, 35] (**Figure 3**).

5.4. Haspin function in mitosis

In vertebrate cell lines, depletion of Haspin by siRNA or treatment with specific inhibitors leads to a substantial decrease in histone H3 Thr3 phosphorylation during mitosis. Moreover,

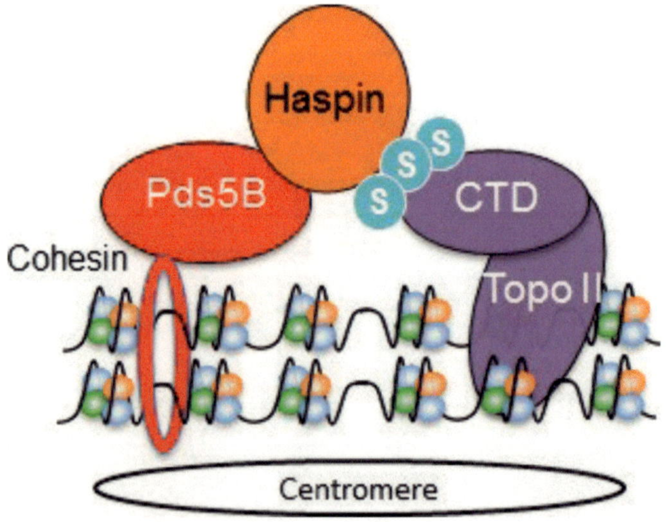

Figure 3. Haspin recruitment at centromere. Haspin centromeric localization is dependent on binding to both the cohesin-associated protein Pds5B and the sumoylated C-terminal domain (CTD) of Topo isomerase II (Topo II).

http://dx.doi.org/10.5772/intechopen.70732

cells display many remarkable defects in mitosis. Prometaphase and metaphase duration are increased due to severe chromosome alignment defects [6, 7, 13] (**Figure 4**).

Mitosis duration, measured by video microscopy, is increased up to several hours in Haspin-depleted U-2 OS cells [36]. Mitotic spindles are disorganized with often extra centrosome-like foci. Impaired centromeric cohesion and premature separation of chromatid have been reported in Higgins *et al.* [36].

Haspin has been shown to be the major H3T3 phosphorylating kinase in various organisms [7, 8, 10, 11, 13, 15]. Phosphorylation on H3T3 appears first at the end of G2 phase of the cell cycle and disappears during anaphase B (**Figure 5** upper panel). H3T3ph is well marked in prophase. At this stage, it is nuclear and more precisely located on condensing chromosome arms. During prometaphase, phosphorylation is concentrated on centromeres in a region delimited by centromeric CENP-A (inner centromere). The phosphorylation decreases rapidly at anaphase and can still be observed on telomeres present in the vicinity of Aurora B activity area on the midzone. It is no longer detected on chromosomes, when cells are in late telophase (**Figure 5** upper panel) [6, 7, 13, 14].

The H3T3ph-dephosphorylating enzyme has been shown to be the PP1γ phosphatase specifically targeted to anaphase chromosomes by its regulatory subunit Repo-Man [37, 38]. Further studies showed that Repo-Man targeting to chromosomes is negatively regulated through Aurora B phosphorylation explaining the persistent H3T3ph signal observed on telomeres at anaphase [37].

Histone H3 phosphorylated on threonine 3 is directly recognized by the conserved BIR domain of Survivin, a member of the chromosomal passenger complex (CPC) [14, 19, 39], thus

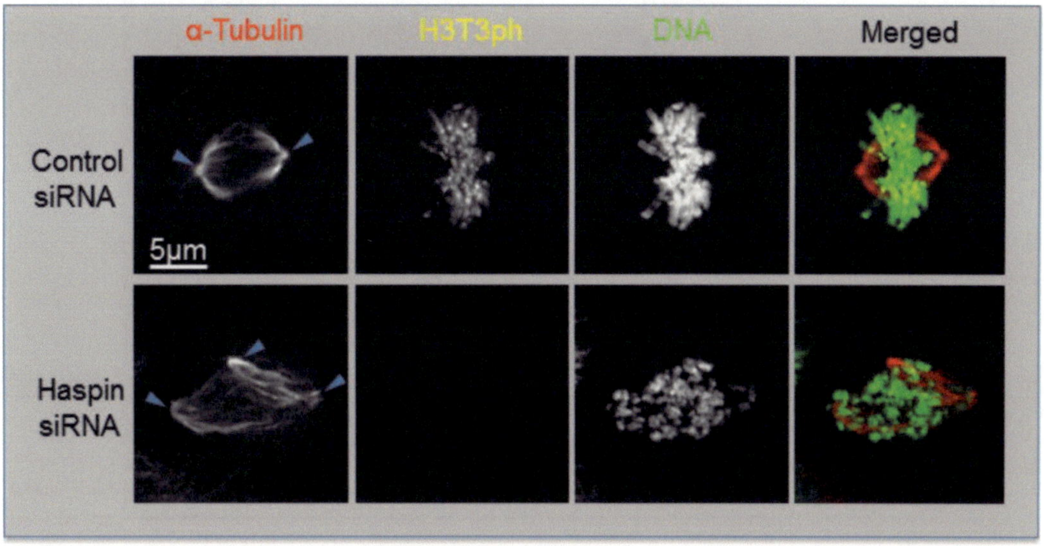

Figure 4. Haspin depletion by siRNA. Immunofluorescence images of Haspin and control siRNA on U2 OS cells. Haspin-depleted cells show chromosome alignment defects, impaired spindles, and ectopic spindle poles.

anchoring the CPC at centromeres. The CPC is a complex of four subunits, Survivin, Borealin, INCENP, and the Aurora B kinase. Aurora B is an essential kinase, which regulates mitotic progression, including spindle assembly checkpoint, condensation, and chromosomal bi-orientation and cytokinesis [40–42]. Additionally, Aurora B phosphorylates Haspin N-terminus

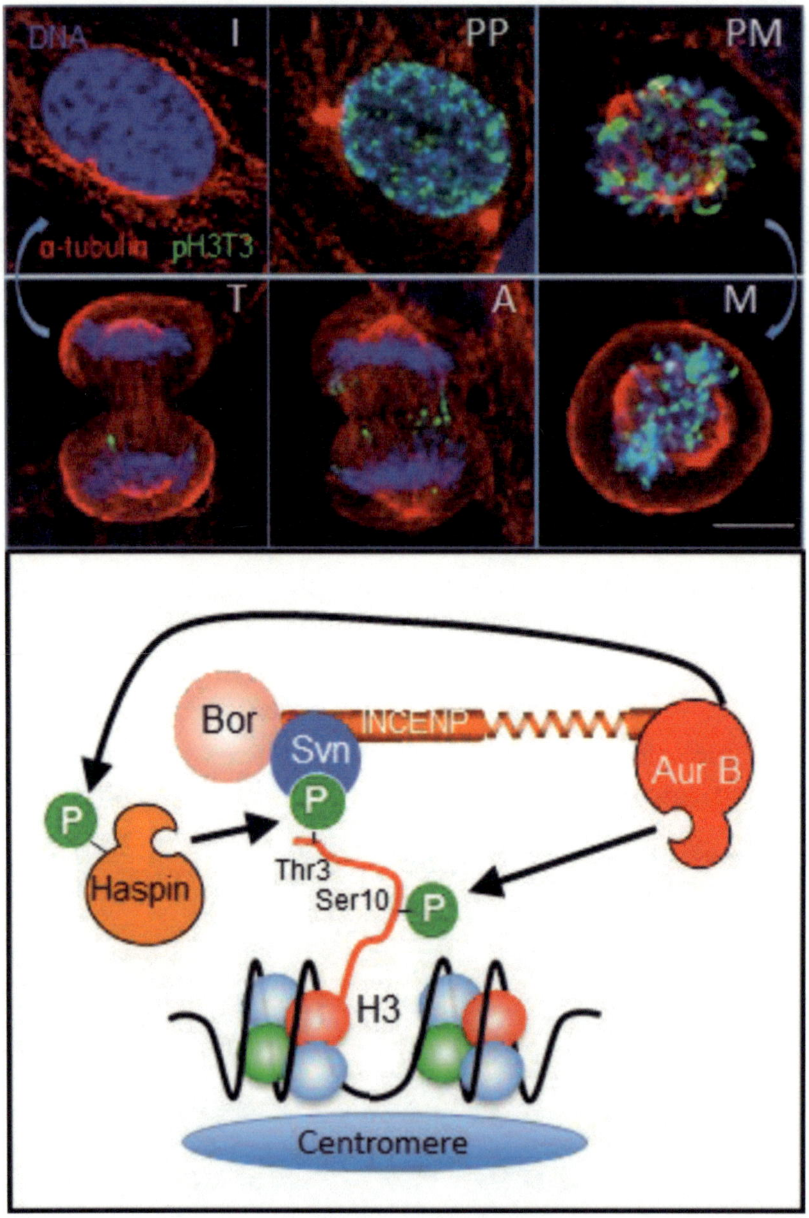

Figure 5. Haspin activity on Thr3 of histone H3. Upper panel, localization of Haspin activity on Thr3 of histone H3 along the cell cycle in mammal cells. Lower panel, schematic representation of Haspin-aurora B positive feedback loop at centromere.

http://dx.doi.org/10.5772/intechopen.70732

on several sites allowing its over-activation and creating a positive feedback loop triggering the accumulation of CPC at the centromeres [14] (**Figure 4** lower panel).

The recruitment of CPC at centromere is not only dependent on H3T3ph by Haspin. A second parallel pathway involving histone H2A phosphorylation on Thr120 by Bub1 creates a binding site for Shugoshin, a protein involved in the protection of centromeric cohesion. Shugoshin binds directly to Survivin in yeast and to Borealin CPC member in human in a comparable manner [14, 43, 44].

Therefore, one of the major functions of Haspin is, together with Bub1 kinase, to bring the CPC at centromeres (**Figure 6**).

Haspin has been shown to be involved in chromosomal cohesion. Defects in chromosome alignment in Haspin-depleted cells are probably due, at least in part, to a premature loss of sister chromatid cohesion [6].

A recent study has demonstrated that, during mitosis, Haspin binds to the cohesin-associated protein Pds5B [45]. During prophase and prometaphase, sister chromatids resolution occurs through cohesion release upon binding of Wapl protein to Pds5B. Zhou *et al.* showed that Haspin interaction with Pds5B inhibits Wapl binding, protecting from premature centromeric cohesion loss [45].

5.5. Regulation of Haspin activity

The N-terminal domain of Haspin appears to be involved in both the intracellular localization of the protein and in the regulation of its kinase domain activity. Indeed, it has been shown that presence of the N-terminal domain changes the phosphorylation kinetics of Histone H3 substrate peptides, when compared to the catalytic C-terminal part alone, increasing the K_m for ATP and lowering the affinity for Histone H3 [10]. Thus, the N-terminal domain has the potential to modulate the activity of the enzyme [10].

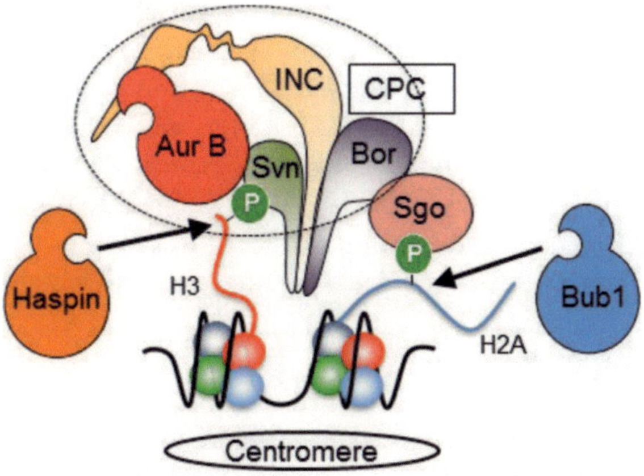

Figure 6. Haspin, together with Bub1, is required for anchoring the CPC at centromere. Svn: Survivin; INC: INCENP; Bor: Borealin; Aur B: Aurora B; Sgo: Shugoshin; CPC: Chromosomal passenger complex.

Haspin is expressed throughout the cell cycle [13, 46]. However, it is highly phosphorylated during mitosis [13]. Phosphoproteomic studies showed that these phosphorylations are on the N-terminal domain of the protein, where phosphorylation consensus sites for Cdk1, Plk1, and Aurora B are present [45–47]. Phosphoryation events on the N-terminal domain at the onset of mitosis trigger conformational changes and influence Haspin kinetics parameters (see above). Haspin phosphorylation by Cdk1/cyclin B starts on T128 of human Haspin (T206 in *Xenopus laevis*) [47, 48]. Gheniou *et al.* showed that Xenopus Haspin auto-inhibits itself during interphase through a conserved basic site in its N-terminus part close to its kinase domain [47]. This auto-inhibition is released through Cdk1 phosphorylation of Haspin N-terminus followed by the recruitment of Plk1 on the Cdk1 phospho-site and its activation. Activated Plk1 phosphorylates several sites on Haspin N-terminus releasing its activity in a timely manner at the beginning of mitosis triggering H3T3 phosphorylation and CPC recruitment at centromeres [47, 48]. Furthermore, Wang *et al.* showed that Aurora B further phosphorylates Haspin N-terminus enhancing its ability to generate H3T3 phospho-sites for Survivin/CPC binding [49] (**Figure 5** lower panel). Another recent study showed that Aurora A also phosphorylates Haspin N-terminus triggering the Aurora B/Haspin feedback loop [50].

Several reports showed that H3T3 phosphorylation by Haspin is regulated by modifications on adjacent residues Arg2 and Lys4. As such, methylation on Arg2 as well as acetylation and methylation on Lys4 strongly decreased the ability of Haspin to phosphorylate Thr3 [8, 51]. These results imply a likely epigenetic regulation of Haspin and Aurora B activities.

5.6. Haspin function in meiosis

Studies of Haspin function during meiosis were performed on mouse oocytes using small molecule inhibitors and overexpression. Similarly to mitosis, Haspin phosphorylates Thr3 of histone H3 [29, 46]. This phosphorylation is necessary for accurate meiosis including chromatin condensation and formation of the microtubule assembly checkpoint, ensuring faithful segregation of chromosomes during meiosis I [29, 46]. Furthermore, Haspin phosphorylation on H3T3 has been shown to be required for Aurora C kinase proper localization during meiosis [29, 46]. During meiosis I, in contrast to mitosis, where Haspin is involved in CPC targeting to centromeres, Haspin regulates Aurora C localization to the inter-chromatid axis [29, 30]. Inhibition of Haspin by 5-ITu showed a failure to organize microtubules and an increase in microtubule organizing centers (MTOCs) as well as an impaired localization of Aurora C at this location. These results suggested a new role for Haspin in the regulation of MTOCs clustering during meiosis and Aurora C localization at MTOC supporting the idea of different functions of Haspin in meiosis compared to mitosis [30].

6. Haspin as a therapeutic target

Mitotic protein kinases are considered as targets of choice for drugs developed by the pharmaceutical industry [52]. Because of its role in controlling the activity of Aurora B and in maintaining the cohesion of centromeres and spindle poles, Haspin has become a relevant target for cancer therapy and is considered as an emerging anti-mitotic drug target [53, 54].

The fact that Haspin is an atypical ePK with a divergent structure may lead to the development of inhibitors with fewer side effects [4, 5]. Haspin inhibitor CHR6494 described by Huertas *et al.* showed antitumor activity in a xenograph mouse model [53]. Haspin is also overexpressed in some malignant tumors such as Burkitt's lymphoma and chronic lymphocytic leukemias [55, 56]. In addition, Haspin was identified in a whole kinome siRNA screen, together with Plk1, as one of the top hit kinases, whose depletion decreased both cell viability and estrogen receptor transcriptional activity in MCF7 breast cancer cells [57]. Thus, Haspin may represent a new anti-cancer therapeutic target.

7. Haspin inhibitors

There are only few reports on conception of Haspin inhibitors (**Figure 7**). Most publications reported the evaluation of molecules on a kinase panel, including Haspin kinase.

One of the first molecules used in researches on Haspin was the well-known 5-ITu for its potent inhibition of adenosine kinase. This nucleotide-like molecule inhibits strongly Haspin with IC_{50} ranged between 5 and 9 nM. Initially, 5-ITu was mostly used to get a better understanding of Haspin structure and could also be considered as a tool for biological studies. 5-ITu was recently used to assess biological function of Haspin on the cell cycle, especially during mitosis and meiosis (see above). Cuny *et al.* described the screening, synthesis, and biological

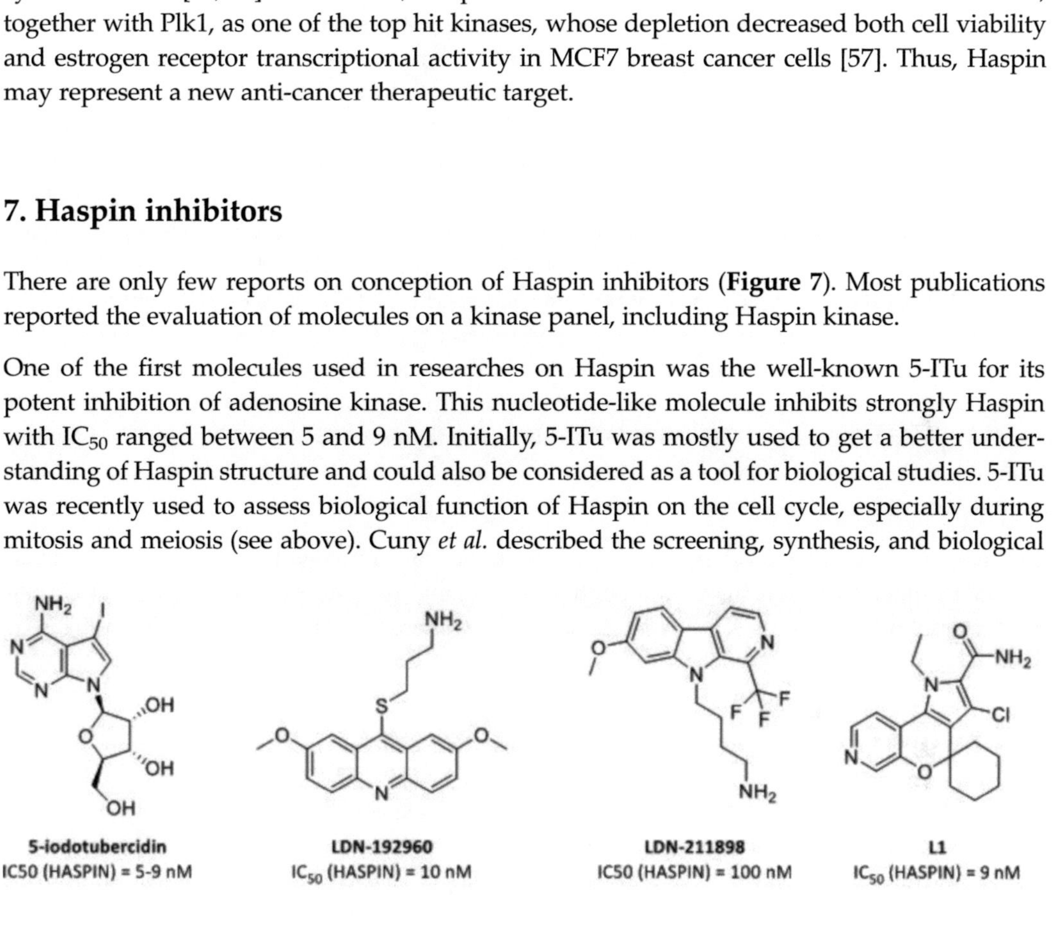

Figure 7. Haspin inhibitors described in the literature.

evaluation of two compounds with interesting activities on Haspin [58]. The study was mostly devoted to target Dyrk2 and Haspin kinases for their role on proliferating cells. After a screening of 140,000 species, an acridine analog demonstrated an interesting profile, and authors isolated LDN-192960, which showed remarkable inhibition of Haspin kinase (IC_{50} = 10 nM). In 2012, the same team synthesized a library of harmine derivative with the same amino-alkyl chain. The newly generated LDN-211898 is described as an inhibitor active at submicromolar concentration against Haspin in an *in vitro* assay (100 nM). Recently, Novartis realized a large screen using Melk inhibitors and found that the compound Melk8a had the best inhibition potency against Haspin (IC_{50} = 190 nM). However, this compound also showed activities below 1 μM of other kinases including Gsk3, Cdk2, Akt1, Flt3 and was therefore not selective. This year, Pastor Fernández *et al.* described the synthesis of tricyclic compounds as new kinases inhibitors [59]. Mostly, the patented molecules have strong activities against Cdk8, Cdk19, and Haspin, as shown for inhibitor L1 (IC_{50} = 9 nM). A similar approach was used by Chen *et al.*, who first described SGI-1776 as a Pim1 inhibitor [60]. *In vitro* evaluation of this compound on a panel of kinases gave an IC_{50} of 34 nM on Haspin. This study was the starting point for screening of other imidazopyridazine as strong Haspin inhibitors. In 2012, Huertas *et al.* described *in vitro,* in cells, and *in vivo* activities of a little imidazoipyridazine named CHR6494 [53]. This molecule seems to be an ATP competitive drug commonly denominated as a type I kinase inhibitor showing a strong inhibition of Haspin with a remarkable IC_{50} of 2 nM.

Kestav *et al.* developed another original type of inhibitors. They synthetized conjugates bearing an aromatic fragment fused to a peptide mimicking the N-terminus of histone H3. Their best compounds showed a K_d of 0.42 nM on Haspin kinase with a good selectivity index [61].

8. Conclusion

Haspin protein kinase was discovered two decades ago. Despite several years of research, the only well characterized substrate, with a specific function, is Histone H3 Thr3. This atypical kinase and its essential role in the regulation of CPC activity in space and time along mitosis, through Thr3 phosphorylation of Histone H3, has become a very attractive subject. The latest findings reviewed here show that there is still much to discover about the function and regulation of this kinase. Although Haspin inhibitors have shown to be very useful tools in dissecting the kinase function in diverse biological mechanisms within multiple organisms, we expect to see their development toward therapeutic drugs in the coming years.

Acknowledgements

We thank Dr. Jonathan Dorival (Roscoff Biological Station, UMR8227) for his help on illustrating **Figure 1** and Dr. Julien Duez (Roscoff Biological Station, USR3151) for helpful criticisms on text and figures. Our work is supported by La Ligue Contre le Cancer, Fondation ARC, Cancéropôle GrandOuest and GIS Biogenouest. OF was supported by a Brittany Region ARED Doctoral fellowship.

http://dx.doi.org/10.5772/intechopen.70732

Author details

Omid Feizbakhsh[1], Matthieu Place[2], Xavier Fant[1], Frédéric Buron[2], Sylvain Routier[2] and Sandrine Ruchaud[1]*

*Address all correspondence to: sruchaud@sb-roscoff.fr

1 Protein Phosphorylation & Human Diseases, Station Biologique de Roscoff, Sorbonne Universités, Roscoff, France

2 Institut de Chimie Organique et Analytique, Université d'Orléans, Orléans, France

References

[1] Ma HT, Poon RYC. How protein kinases co-ordinate mitosis in animal cells. The Biochemical Journal. 2011 Apr;**435**(1):17-31

[2] Combes G, Alharbi I, Braga LG, Elowe S. Playing polo during mitosis: PLK1 takes the lead. Oncogene [Internet]. 2017 Apr;24 [cited 2017 Jul 26]; Available from: http://www.nature.com.insb.bib.cnrs.fr/onc/journal/vaop/ncurrent/full/onc2017113a.html?foxtrotcallback=true

[3] Tanaka H, Yoshimura Y, Nishina Y, Nozaki M, Nojima H, Nishimune Y. Isolation and characterization of cDNA clones specifically expressed in testicular germ cells. FEBS Letters. 1994 Nov 21;**355**(1):4-10

[4] Tanaka H, Yoshimura Y, Nozaki M, Yomogida K, Tsuchida J, Tosaka Y, et al. Identification and characterization of a haploid germ cell-specific nuclear protein kinase (Haspin) in spermatid nuclei and its effects on somatic cells. The Journal of Biological Chemistry. 1999 Jun 11;**274**(24):17049-17057

[5] Higgins JMG. Haspin-like proteins: A new family of evolutionarily conserved putative eukaryotic protein kinases. Protein Science. 2001 Aug 1;**10**(8):1677-1684

[6] Dai J, Sullivan BA, Higgins JMG. Regulation of mitotic chromosome cohesion by Haspin and aurora B. Developmental Cell. 2006 Nov 1;**11**(5):741-750

[7] Markaki Y, Christogianni A, Politou AS, Georgatos SD. Phosphorylation of histone H3 at Thr3 is part of a combinatorial pattern that marks and configures mitotic chromatin. Journal of Cell Science. 2009 Aug 15;**122**(16):2809-2819

[8] Eswaran J, Patnaik D, Filippakopoulos P, Wang F, Stein RL, Murray JW, et al. Structure and functional characterization of the atypical human kinase haspin. Proceedings of the National Academy of Sciences. 2009 Dec 1;**106**(48):20198-20203

[9] Higgins JMG. Structure, function and evolution of haspin and haspinrelated proteins, a distinctive group of eukaryotic protein kinases. Cellular and Molecular Life Sciences. 2003 Mar 1;**60**(3):446-462

[10] Villa F, Capasso P, Tortorici M, Forneris F, de MA, Mattevi A, et al. Crystal structure of the catalytic domain of Haspin, an atypical kinase implicated in chromatin organization. Proceedings of the National Academy of Sciences. 2009 Dec 1;**106**(48):20204-20209

[11] Maiolica A, de M-RM, Schoof EM, Chaikuad A, Villa F, Gatti M, et al. Modulation of the chromatin Phosphoproteome by the Haspin protein kinase. Molecular & Cellular Proteomics. 2014 Jul 1;**13**(7):1724-1740

[12] Huse M, Kuriyan J. The conformational plasticity of protein kinases. Cell. 3 May 2002; **109**(3):275-282

[13] Dai J, Sultan S, Taylor SS, Higgins JMG. The kinase haspin is required for mitotic histone H3 Thr 3 phosphorylation and normal metaphase chromosome alignment. Genes & Development. 2005 Feb 15;**19**(4):472-488

[14] Wang F, Ulyanova NP, van der Waal MS, Patnaik D, Lens SMA, Higgins JMG. A positive feedback loop involving Haspin and aurora B promotes CPC accumulation at centromeres in mitosis. Current Biology. 2011 Jun 21;**21**(12):1061-1069

[15] Polioudaki H, Markaki Y, Kourmouli N, Dialynas G, Theodoropoulos PA, Singh PB, et al. Mitotic phosphorylation of histone H3 at threonine 3. FEBS Letters. 2004 Feb 27;**560**(1–3):39-44

[16] Kozgunova E, Suzuki T, Ito M, Higashiyama T, Kurihara D. Haspin has multiple functions in the plant cell division regulatory network. Plant & Cell Physiology. 2016 Apr;**57**(4):848-861

[17] Ashtiyani RK, Moghaddam AMB, Schubert V, Rutten T, Fuchs J, Demidov D, et al. AtHaspin phosphorylates histone H3 at threonine 3 during mitosis and contributes to embryonic patterning in Arabidopsis. The Plant Journal. 2011 Nov 1;**68**(3):443-454

[18] Panigada D, Grianti P, Nespoli A, Rotondo G, Gallo Castro D, Quadri R, et al. Yeast Haspin kinase regulates polarity cues necessary for mitotic spindle positioning and is required to tolerate mitotic arrest. Developmental Cell. 2013 Sep 16;**26**(5):483-495

[19] Yamagishi Y, Honda T, Tanno Y, Watanabe Y. Two histone marks establish the inner centromere and chromosome bi-orientation. Science. 2010 Oct 8;**330**(6001):239-243

[20] Costanzi C, Pehrson JR. Histone macroH2A1 is concentrated in the inactive X chromosome of female mammals. Nature. 1998 Jun 11;**393**(6685):599-601

[21] Kim W, Chakraborty G, Kim S, Shin J, Park C-H, Jeong M-W, et al. Macro histone H2A1.2 (MacroH2A1) protein suppresses mitotic kinase VRK1 during interphase. The Journal of Biological Chemistry. 2012 Feb 17;**287**(8):5278-5289

[22] Muthurajan UM, Bao Y, Forsberg LJ, Edayathumangalam RS, Dyer PN, White CL, et al. Crystal structures of histone sin mutant nucleosomes reveal altered protein–DNA interactions. The EMBO Journal. 2004 Jan 28;**23**(2):260-271

[23] Gascoigne KE, Takeuchi K, Suzuki A, Hori T, Fukagawa T, Cheeseman IM. Induced ectopic kinetochore assembly bypasses the requirement for CENP-A nucleosomes. Cell. 2011 Apr 29;**145**(3):410-422

[24] Kimura M, Kotani S, Hattori T, Sumi N, Yoshioka T, Todokoro K, et al. Cell cycle-dependent expression and spindle pole localization of a novel human protein kinase, Aik, related to aurora of drosophila and yeast Ipl1. The Journal of Biological Chemistry. 23 May 1997;**272**(21):13766-13771

[25] Golsteyn RM, Mundt KE, Fry AM, Nigg EA. Cell cycle regulation of the activity and subcellular localization of Plk1, a human protein kinase implicated in mitotic spindle function. 1995. [Jun 15 [cited 17 May 2017]] Available from: https://lra.le.ac.uk/handle/2381/38961

[26] Hindriksen S, Bramer AJ, Truong MA, Vromans MJM, Post JB, Verlaan-Klink I, et al. Baculoviral delivery of CRISPR/Cas9 facilitates efficient genome editing in human cells. PLoS One [Internet]. 2017 Jun 22 [cited 2017 Jul 13];**12**(6): Available from: http://www.ncbi.nlm.nih.gov/pmc/articles/PMC5480884/

[27] Caperta AD, Rosa M, Delgado M, Karimi R, Demidov D, Viegas W, et al. Distribution patterns of phosphorylated Thr 3 and Thr 32 of histone H3 in plant mitosis and meiosis. Cytogenetic and Genome Research. 2008;**122**(1):73-79

[28] Houben A, Demidov D, Caperta AD, Karimi R, Agueci F, Vlasenko L. Phosphorylation of histone H3 in plants—A dynamic affair. Biochimica et Biophysica Acta (BBA): Gene Structure and Expression. 2007 May;**1769**(5–6):308-315

[29] Nguyen AL, Gentilello AS, Balboula AZ, Shrivastava V, Ohring J, Schindler K. Phosphorylation of threonine 3 on histone H3 by haspin kinase is required for meiosis I in mouse oocytes. Journal of Cell Science. 2014 Dec 1;**127**(23):5066-5078

[30] Balboula AZ, Nguyen AL, Gentilello AS, Quartuccio SM, Drutovic D, Solc P, et al. Haspin kinase regulates microtubule-organizing center clustering and stability through aurora kinase C in mouse oocytes. Journal of Cell Science. 2016 Oct 1;**129**(19):3648-3660

[31] Losada A, Yokochi T, Hirano T. Functional contribution of Pds5 to cohesin-mediated cohesion in human cells and *Xenopus* egg extracts. Journal of Cell Science. 15 May 2005;**118**(10):2133-2141

[32] Sumara I, Vorlaufer E, Gieffers C, Peters BH, Peters J-M. Characterization of vertebrate cohesin complexes and their regulation in prophase. The Journal of Cell Biology. 2000 Nov 13;**151**(4):749-762

[33] Carretero M, Ruiz-Torres M, Rodríguez-Corsino M, Barthelemy I, Losada A. Pds5B is required for cohesion establishment and aurora B accumulation at centromeres. The EMBO Journal. 2013 Nov 13;**32**(22):2938-2949

[34] Yoshida MM, Ting L, Gygi SP, Azuma Y. SUMOylation of DNA topoisomerase IIα regulates histone H3 kinase Haspin and H3 phosphorylation in mitosis. The Journal of Cell Biology. 2016 Jun 20;**213**(6):665-678

[35] Edgerton H, Johansson M, Keifenheim D, Mukherjee S, Chacón JM, Bachant J, et al. A noncatalytic function of the topoisomerase II CTD in aurora B recruitment to inner centromeres during mitosis. The Journal of Cell Biology. 2016 Jun 20;**213**(6):651-664

[36] Dai J, Kateneva AV, Higgins JMG. Studies of haspin-depleted cells reveal that spindle-pole integrity in mitosis requires chromosome cohesion. Journal of Cell Science. 2009 Nov 15;**122**(22):4168-4176

[37] Qian J, Lesage B, Beullens M, Van Eynde A, Bollen M. PP1/repo-man dephosphorylates mitotic histone H3 at T3 and regulates chromosomal aurora B targeting. Current Biology. 10 May 2011;**21**(9):766-773

[38] Vagnarelli P, Ribeiro S, Sennels L, Sanchez-Pulido L, de Lima Alves F, Verheyen T, et al. Repo-man coordinates chromosomal reorganization with nuclear envelope reassembly during mitotic exit. Developmental Cell. 2011 Aug 16;**21**(2):328-342

[39] Kelly AE, Ghenoiu C, Xue JZ, Zierhut C, Kimura H, Funabiki H. Survivin reads phosphorylated histone H3 threonine 3 to activate the mitotic kinase aurora B. Science. 2010 Oct 8;**330**(6001):235-239

[40] Carmena M, Wheelock M, Funabiki H, Earnshaw WC. The chromosomal passenger complex (CPC): From easy rider to the godfather of mitosis. Nature Reviews. Molecular Cell Biology. 2012 Dec;**13**(12):789-803

[41] van der Horst A, Lens SMA. Cell division: Control of the chromosomal passenger complex in time and space. Chromosoma. 2014;**123**(1–2):25-42

[42] Ruchaud S, Carmena M, Earnshaw WC. Chromosomal passengers: Conducting cell division. Nature Reviews. Molecular Cell Biology. 2007 Oct;**8**(10):798-812

[43] Lordier L, Chang Y, Jalil A, Aurade F, Garçon L, Lécluse Y, et al. Aurora B is dispensable for megakaryocyte polyploidization, but contributes to the endomitotic process. Blood. 2010 Sep 30;**116**(13):2345-2355

[44] Storchová Z, Becker JS, Talarek N, Kögelsberger S, Pellman D. Bub1, Sgo1, and Mps1 mediate a distinct pathway for chromosome biorientation in budding yeast. Molecular Biology of the Cell. 1 May 2011;**22**(9):1473-1485

[45] Zhou L, Liang C, Chen Q, Zhang Z, Zhang B, Yan H, et al. The N-terminal non-kinase-domain-mediated binding of Haspin to Pds5B protects Centromeric cohesion in mitosis. Current Biology. 2017 Apr 3;**27**(7):992-1004

[46] Wang Q, Wei H, Du J, Cao Y, Zhang N, Liu X, et al. H3 Thr3 phosphorylation is crucial for meiotic resumption and anaphase onset in oocyte meiosis. Cell Cycle. 2015 Dec 4;**15**(2):213-224

[47] Ghenoiu C, Wheelock MS, Funabiki H. Autoinhibition and polo-dependent multisite phosphorylation restrict activity of the histone H3 kinase Haspin to mitosis. Molecular Cell [Internet]. 2013 Dec 12 [cited 2017 Jul 13];**52**(5): Available from: http://www.ncbi.nlm.nih.gov/pmc/articles/PMC3865225/

[48] Zhou L, Tian X, Zhu C, Wang F, Higgins JM. Polo-like kinase-1 triggers histone phosphorylation by Haspin in mitosis. EMBO Reports. 2014 Mar;**15**(3):273-281

[49] Wang F, Dai J, Daum JR, Niedzialkowska E, Banerjee B, Stukenberg PT, et al. Histone H3 Thr-3 phosphorylation by Haspin positions aurora B at centromeres in mitosis. Science. 2010 Oct 8;**330**(6001):231-235

[50] Yu F, Jiang Y, Lu L, Cao M, Qiao Y, Liu X, et al. Aurora-a promotes the establishment of spindle assembly checkpoint by priming the Haspin-aurora-B feedback loop in late G2 phase. Cell Discovery. 2017 Jan 10;**3**:celldisc201649

[51] Han A, Lee KH, Hyun S, Lee NJ, Lee SJ, Hwang H, et al. Methylation-mediated control of aurora kinase B and Haspin with epigenetically modified histone H3 N-terminal peptides. Bioorganic & Medicinal Chemistry. 2011 Apr 1;**19**(7):2373-2377

[52] de Cárcer G, Pérez de Castro I, Malumbres M. Targeting Cell Cycle Kinases for Cancer Therapy [Internet]. http://www.eurekaselect.com. Available from: http://www.eurekaselect.com/58964/article

[53] Huerta EE. Cancer prevention in the 21st century: The Preventorium concept. Journal of Health Care for the Poor and Underserved. 2012 Oct 31;**23**(4):1698-1703

[54] Dominguez-Brauer C, Thu KL, Mason JM, Blaser H, Bray MR, Mak TW. Targeting mitosis in cancer: Emerging strategies. Molecular Cell. 2015 Nov 19;**60**(4):524-536

[55] Rosenwald A, Alizadeh AA, Widhopf G, Simon R, Davis RE, Yu X, et al. Relation of gene expression phenotype to immunoglobulin mutation genotype in B cell chronic lymphocytic Leukemia. The Journal of Experimental Medicine. 2001 Dec 3;**194**(11):1639-1648

[56] Dave SS, Fu K, Wright GW, Lam LT, Kluin P, Boerma E-J, et al. Molecular diagnosis of Burkitt's lymphoma. The New England Journal of Medicine. 2006 Jun 8;**354**(23):2431-2442

[57] Bhola NE, Jansen VM, Bafna S, Giltnane JM, Balko JM, Estrada MV, et al. Kinome-wide functional screen identifies role of PLK1 in hormone-independent, ER-positive breast cancer. Cancer Research. 2015 Jan 15;**75**(2):405-414

[58] Cuny GD, Ulyanova NP, Patnaik D, Liu J-F, Lin X, Auerbach K, et al. Structure–activity relationship study of beta-carboline derivatives as haspin kinase inhibitors. Bioorganic & Medicinal Chemistry Letters. 2012 Mar 1;**22**(5):2015-2019

[59] Brevet US9682991. Tricyclic compounds for use as kinase inhibitors. Google Brevets [Internet] [cited 2017 Jul 27]. Available from: https://www.google.com/patents/US9682991

[60] Chen LS, Redkar S, Bearss D, Wierda WG, Gandhi V. Pim kinase inhibitor, SGI-1776, induces apoptosis in chronic lymphocytic leukemia cells. Blood. 2009 Nov 5;**114**(19):4150-4157

[61] Kestav K, Lavogina D, Raidaru G, Chaikuad A, Knapp S, Uri A. Bisubstrate inhibitor approach for targeting mitotic kinase Haspin. Bioconjugate Chemistry. 2015 Feb 18;**26**(2):225-234

Chapter 3

Regulation of Aurora Kinases and Their Activity

Lucie Vaufrey, Arun Prasath Damodaran,
Olivia Gavard, Stéphanie Le Bras and Claude Prigent

Additional information is available at the end of the chapter

http://dx.doi.org/10.5772/intechopen.70595

Abstract

The three mitotic protein kinases Aurora-A, B and C are complementary enzymes that regulate multiple mitotic events. To do so, the different kinases must be locally activated, and the control of their activity is tightly regulated in time and space during mitosis. For instance, Aurora-A is first active at the centrosomes, then on microtubules at the spindle pole, Aurora-B is active in the nucleus, then at chromosome kinetochores and later one at the midbody. Aurora kinase activity is regulated in space and time by locally binding to regulators. Aurora kinases must bind to protein partners to be activated. Aurora-A for instance binds to targeting protein for *Xenopus* kinesin-like protein 2 (TPX2) and is activated at the spindle pole, Aurora-B and Aurora-C to INner CENtromer Protein (INCENP) and is activated on the chromosomes. These activations go through an autophosphorylation of a threonine residue in the T-loop of the kinase. Other protein partners are using different mechanisms to activate Auroras. These allow activation of the kinase at different time and location in the cell. This review is an up-to-date list of regulators of Aurora kinases. The subcellular localization of these regulators explains the presence of an active Aurora kinase. It also explains the changes in the localizations of the Aurora kinases activity observed during cell cycle progression in mitosis. Aurora kinases have been recently reported to be involved in nonmitotic events, and the identity of their activators in these events must be searched.

Keywords: Aurora kinase, regulation, posttranslational modification

1. Introduction

During the process of division, the cell goes through two main phases such as interphase and mitosis that is followed by the physical separation of the two daughter cells. During interphase, the cell duplicates its contents that will be segregated during mitosis to generate two daughter cells. The whole process lasts about 24 h in the case of human cells during which mitosis takes only 1 h. This short phase is highly regulated by phosphorylation and dephosphorylation reactions [1]. Among the key protein kinases involved, there are cyclin-dependent kinase 1 (CDK1), polo-like kinase 1 (Plk1), NIMA-related kinase 2 (Nek2) and the Aurora kinases (Aurora-A, B and C) [2]. The three mitotic protein kinases Aurora-A, B and C are complementary enzymes that regulate multiple mitotic events [3].

During mitosis, the cell segregates its two centrosomes that migrate around the nucleus to reach opposite position. The nuclear membrane breaks down, and the chromatin starts to condense to form chromosomes. Microtubules nucleate from both the centrosomes and the chromosomes to form a bipolar spindle [4]. The force exerted by the spindle microtubules contributes to the alignment of chromosomes on the metaphase plate. This event is immediately followed by the separation of each pair of sister chromatids and the beginning of their migration to the two opposite poles of the cell. The last part of the migration is driven by the central spindle, assembled at the future site of cell division. A constriction ring is assembled around the cell at the exact same location that will contribute to the separation of two volumes corresponding to the two daughter cells [5]. The physical separation of the two cells, abscission, will occur later on during the following interphase.

Aurora-A localizes at both the centrosomes and the spindle poles. The kinase activity is required for microtubule nucleation during bipolar spindle assembly and during central spindle formation [6]. Aurora-B is part of the Chromosome Passenger Complex (CPC), it localizes at chromosome kinetochores from prophase to metaphase and at the midbody from anaphase to telophase [7]. Its kinase activity is responsible for the massive phosphorylation of the Ser10 of histone H3 in mitosis. Aurora-B corrects the wrong attachments of microtubules to kinetochores during prometaphase [8]. During exit from mitosis, Aurora-B is required for cytokinesis. Aurora-C that is mainly involved in meiosis can replace Aurora-B during mitosis [9].

In vitro, the three kinases share the same substrates, such as histone H3. *In vivo,* to fulfill their function, the three kinases are differentially localized and also locally activated, and the control of their activity is then tightly regulated in time and space during mitosis. For instance, Aurora-A is initially active in the cytoplasm and at the centrosomes in the end of G2, then on microtubules at the spindle pole during prometaphase. Aurora-B is active in the nucleus by the end of G2, then at chromosome kinetochores in prometaphase and later at the midbody during anaphase and telophase. To achieve such regulation in space and time Aurora kinases are locally activated by binding to activators.

2. Bipolar spindle assembly

2.1. Aurora-A and TPX2

Targeting protein for *Xenopus* kinesin-like protein 2 (TPX2) is the best-characterized Aurora-A activator. It is a 100-kDa protein expressed from G1/S transition to cytokinesis and then rapidly degraded [10, 11]. TPX2 was first identified as a binding partner of the plus end-directed

Xenopus kinesin-like protein 2 (Xklp2) [12]. TPX2 helps localizing Xklp2 to the spindle pole in prometaphase and metaphase [13, 14].

In interphase, from S to G2, TPX2 localizes to the nucleus where it is sequestered by importin alpha. In mitosis, TPX2 is released from the importin by RanGTP in the vicinity of the spindle. RanGTP produced by the chromosome protein RCC1 (RanGEF) is localized as a gradient around the chromosomes. It is only when the nuclear membrane breaks down that TPX2 can reach the centrosome to bind Aurora-A, to activate and re-localize Aurora-A protein on microtubules at the spindle pole [15–17].

The binding of TPX2 to Aurora-A induces a conformational change in the kinase in a way that the phosphorylated threonine of the activation loop (T288 in human) is better protected from phosphatase activity (PP1 in particular) [18]. Phosphorylation of T288 or binding to TPX2 triggers activation of Aurora-A activity independently, but both the events are synergistic [19]. Most importantly, because binding of TPX2 to Aurora-A changes the conformation of the kinase, it modifies the affinity of the substrates for the kinase as well as the affinity of kinase inhibitors [20].

2.2. Aurora-A and CEP192

Centrosomal protein of 192 kDa (Cep192) was named after a proteomic analysis of the centrosome composition [21]. Cep192 is a protein involved in centrosome maturation and bipolar spindle assembly [22]. These functions correspond to those described for Aurora-A [23, 24], and indeed, Cep192 activates Aurora-A at the centrosome to control mitotic spindle assembly [25]. The mechanism by which Cep192 activates Aurora-A is different from TPX2. Cep192 is a scaffold protein that brings together two molecules of Aurora-A. Within the dimer, each Aurora-A molecule phosphorylates its neighbor on T288, leading to the complete activation of the kinase. This activation takes place at the centrosome early in mitosis. Aurora-A is then presumably released from Cep192 to bind to TPX2 and to move on the spindle poles [25].

2.3. Aurora-A and Maskin/TACC3

Maskin is a *Xenopus laevis* protein that got its name by the fact that it regulates RNA translation. Maskin links the 5′cap and the 3′UTR of the mRNA, creating a closed structure that cannot be translated [26]. During *Xenopus laevis* oocyte maturation, phosphorylation of Maskin by Aurora-A is required for the control of sequential mRNA translation [27]. Maskin is not only a substrate of Aurora-A, but it is also an activator of its kinase activity as binding of Maskin to Aurora-A induces a sevenfold stimulation of its kinase activity [27]. Whether Maskin affects the phosphorylation of threonine in the activation loop of the kinase has not been investigated yet.

The human homologue of Maskin is transforming acidic coiled-coil containing protein 3 (TACC3) and D-TACC in *Drosophila melanogaster*. Phosphorylation of D-TACC or TACC3 by Aurora-A is required for its centrosome localization [28], microtubule nucleation during bipolar spindle assembly [29–31] and during central spindle assembly [32]. Conversely, the activation of Aurora-A by TACC3 or D-TACC has not been demonstrated yet.

2.4. Aurora-B and INCENP

INner CENtromer Protein (INCENP) participates to the chromosome passenger complex (CPC) together with Aurora-B, Survivin and Borealin [33]. The complex controls multiple events during mitosis: from chromosome condensation and segregation to cytokinesis [34]. Aurora-B is carrying the kinase catalytic activity of the CPC, while INCENP is the activator of the kinase.

Binding of INCENP to Aurora-B is essential to the function of the kinase such as chromosome segregation and cytokinesis [35, 36]. Just like binding to TPX2 triggers activation of Aurora-A through autophosphorylation of T288 in its activation loop, binding to INCENP triggers activation of Aurora-B through autophosphorylation of T232 in its activation loop [37, 38]. Both the kinases and the modes of activation are so closed that a single amino acid change (G198 to N) transforms the activator of Aurora-A from TPX2 to INCENP. The demonstration has been made in *Xenopus* [39] and human [40].

In term of evolution, it is striking to note that *Drosophila melanogaster* genome do not contain any gene coding for a TPX2. Ssp1/Mei-38 would be the closest TPX2-related protein in *Drosophila*. Ssp1/Mei-38 possesses a microtubule-binding domain but lacks the Aurora-A-binding domain, indicating that it cannot activate Aurora-A [41].

3. G2/M transition

3.1. Aurora-A and Ajuba

Ajuba means curiosity in Urdu, an Indian dialect. Ajuba is a LIM domain-containing protein that serves as a scaffold to build numerous protein complexes. The LIM domain is a Zinc finger structure [42]. Ajuba was first reported to bind to Aurora-A at the centrosome in late G2 and to trigger the kinase activation through autophosphorylation of T288. Ajuba would then participate to the activation of Aurora-A and the commitment to mitosis (**Figure 1**) [43] . It was also suggested that Ajuba interacts with the N-terminal domain of Aurora-A to release its inhibitory binding to the C-terminal catalytic domain of the kinase [43]. Aurora-A activation by Ajuba would be a two-step mechanism, binding to the N-terminal domain of the kinase and triggering autophosphorylation of T288 [44]. The activation of Aurora-A by Ajuba has not been observed In *Xenopus laevis* [45]. In *Drosophila melanogaster*, although Ajuba does not activate Aurora-A, the protein is necessary to maintain Aurora-A at the centrosome [46].

3.2. Aurora-A and Nucleophosmin

Nucleophosmin (NPM) is a nucleolar protein involved in multiple functions: histone chaperoning, ribosome biogenesis, mitotic spindle assembly, genome stability, apoptosis and cancer [47]. Like Aurora-A, NPM localizes to the centrosome where it is required for centrosome duplication [48]. Depletion of NPM leads to the formation of disorganized spindles, a phenotype observed after Aurora-A depletion [49]. NPM is also a strong interactor of Aurora-A, and

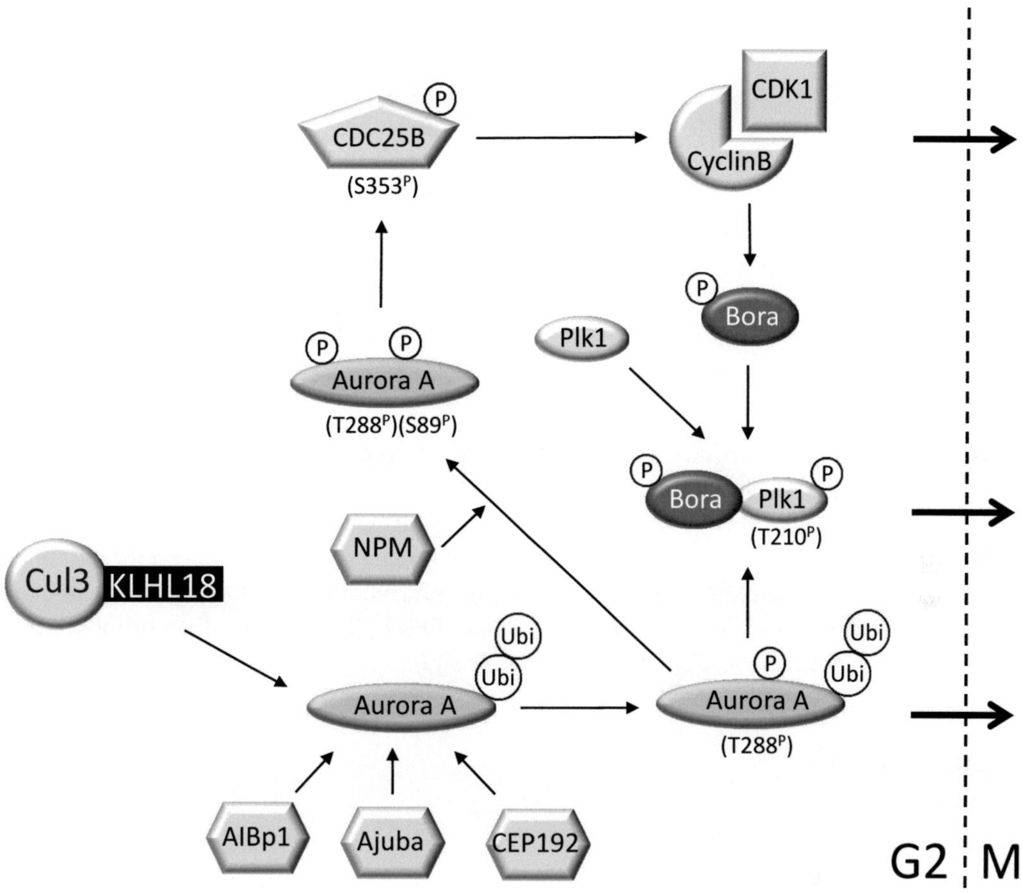

Figure 1. The three kinases CDK1/cyclinB, Plk1 and Aurora-A phosphorylate substrates required the G2/M transition. This scheme shows the pathways used by Aurora-A to activate Plk1 and CDK1/cyclinB.

both proteins interact at the centrosome late in G2. Binding of NPM triggers a phosphorylation event on the kinase. The active Aurora-A, already phosphorylated on T288, undergoes a second autophosphorylation on serine 89 which induces a very strong stimulation of its kinase activity [50]. This stimulation is required at the centrosome in particular for the phosphorylation of S353 on the phosphatase CDC25B involved in the activation of CDK1/cylin B for G2/M transition. Surprisingly, the stimulation of Aurora-A by NPM is not required for the phosphorylation of T210 that activates PLK1 in the end of G2 [50] (**Figure 1**).

3.3. Aurora-A and Bora

Bora was identified in a genetic screen setup to search for mutations affecting the development of *Drosophila melanogaster* external sensory (ES) organs [51]. The gene was named Bora for Aurora-A Borealis because the phenotypes of *Bora* and *Aurora-A* mutants were similar. Bora binds to Aurora-A *in vitro*, is phosphorylated but also activates the kinase in *Drosophila* and human. *In vitro*, Bora can activate Aurora-A in the presence of PP1 (seven- to eightfold), suggesting that the mechanism used by Bora might be identical to TPX2 although it was not demonstrated that Bora triggers autophosphorylation of Aurora-A on threonine in the kinase activation loop [51]. However, when expressed at physiological levels, Bora does not co-immunoprecipitate with Aurora-A, and on the contrary, it immunoprecipitates with PLK1. Furthermore, depletion of Bora does not affect phosphorylation of T288 [52]. Eventually, it was demonstrated that Bora binds to the Polo-Box Domain of Plk1 (PBD) to relieve the auto-inhibition of PBD and to expose the T210 of the activation loop to Plk1-activating kinase. Aurora-A then binds to Bora, gets activated and activates Plk1 by phosphorylating T210 (**Figure 1**) [52, 53]. This activation of Plk1 by Aurora-A through the interaction with Bora occurs in G2.

3.4. Aurora-A and AIBp1

AIBp1 (AIK binding protein, AIK stands for Aurora/Ipl1-related kinase) is thus an Aurora-A binding protein but also a hNinein binding protein. Depletion of AIBp1 gives phenotypes typical of Aurora-A: bipolar mitotic spindle defects [54]. Binding of AIBp1 to Aurora-A increases its kinase activity *in vitro* [54]. *In vivo* expression of AIBp1 increases T288 phosphorylation on Aurora-A, whereas its depletion decreases T210 phosphorylation on Plk1 [55]. These data are reminiscent of the effect of Bora on both Aurora-A and Plk1 [52, 53]. It was then proposed that AIBp1 plays the same role as Bora but in a hNinein signaling pathway.

4. Actin network

4.1. Aurora-A and HEF-1/NEDD9/Cas-L

Human enhancer of filamentation 1 (HEF1) or neural precursor cell expressed, developmentally down-regulated 9 (NEDD9) or Crk-associated substrate related, lymphocyte-type (Cas-L) is a scaffolding protein that localizes to focal adhesions in interphase cells and to the mitotic spindle in M-phase. It participates in integrin-dependent signaling processes, such as cell attachments, cell migration and cell survival [56]. Cells depleted with HEF-1 show a decrease in T288 phosphorylation of Aurora-A, indicating that HEF-1 is required for activation of Aurora-A kinase *in vivo*. *In vitro* both proteins directly interact, and when increasing levels of HEF-1 are added to Aurora-A, an increase of T288 phosphorylation and its kinase activity are observed [57]. Interaction of HEF-1 with Aurora-A occurs in G2, during which the activation of Aurora-A by HEF-1 induces phosphorylation of HEF-1 and inhibition of the interaction. Interaction of HEF-1 with Aurora-A plays a critical role in primary cilia disassembly upon reentry in the cell cycle after Go arrest. In this particular case, HEF-1-activated Aurora-A phosphorylates and activates HDAC6, which in turn deacetylates the tubulin that is sufficient to provoke cilia resorption [58]. The activation of Aurora-A in the process of cilia disassembly is also dependent on Ca2+ and calmodulin (CaM) that are required for Aurora-A to bind to its activators [59].

4.2. Aurora-A and PAK-1

p21-Activated protein kinase-1 (PAK-1) regulates cell motility and morphology [60, 61] and is involved in focal adhesion disassembly through the PAK-PIX-GIT complex, PIX is a Rac

GTP exchange factor and GIT is a G-protein-coupled receptor kinase-interacting protein [62]. This complex is also active at the centrosome, and when Pak1 is activated it dissociâtes from the PIX-GIT to phosphorylate and activate Aurora-A [63]. The activation goes through phosphorylation of the T288 in the activation loop, but there was also a phosphorylation of S342 in the C terminal end of the kinase. Although T288 is known to be an activation site, S342 is rather known to inhibit Aurora-A kinase activity when phosphorylated. This has been shown in *Xenopus laevis* where the phosphorylation of S349 (human S342) downregulates Aurora-A *in vitro* [64] and *in vivo* during oocyte maturation between Metaphase I and Metaphase II of meiosis [65]. The same data have been reported in human where phosphorylation of S342 was observed in G2 upon DNA damage to inhibit Aurora-A, avoiding mitosis entry in the presence of lesions [66]. This discrepancy has not been solved yet.

4.3. Aurora-A and ILK

Integrin-linked kinase (ILK), like PKA-1, is a protein kinase involved in cell adhesion, and the kinase links the extracellular matrix to the actin cytoskeleton [67, 68]. ILK has also been observed in centrosome where it associates with TACC3/Ch-TOG, and its kinase activity is required for Aurora-A interaction with TACC3 [69]. ILK acts upstream of Aurora-A that in turn phosphorylates TACC3 on S558, to control microtubule nucleation [70, 71]. How ILK controls Aurora-A activity toward TACC3 is unknown.

4.4. Aurora-A and Arpc1b

Arpc1b is a component of the seven-subunit protein Arp2/3 complex involved in new actin filament nucleation and polymerization [72]. Arpc1b localizes on centrosome in G2 and interacts with Aurora-A only if Arpc1b has been previously phosphorylated on T21 by Pak-1. Arpc1b is also a substrate of Aurora-A, and the kinase phosphorylates wild-type Arpc1b but not the T21A mutant. This phosphorylation by Aurora-A is required for the interaction of Arpc1b with Arpc2 [73]. On the other hand, Arpc1b is an activator of Aurora-A *in vitro* and *in vivo*, and binding to Arpc1b triggers T288 autophosphorylation just like TPX2 does. Depletion of Arpc1b leads to a decrease on T288 as well as a decrease of its activity toward substrates such as PLK1 or histone H3 [73].

5. Ubiquitylation

Ubiquitination corresponds to a posttranslational modification (PTM) of proteins during which 76 amino-acid peptides are covalently linked to a protein, usually on lysine residues. It requires a multistep reaction: it needs 1) an E1 enzyme that will activate the ubiquitin, then 2) an E2 enzyme that will conjugate the ubiquitin and finally 3) an E3-ligase that will catalyze the transfer of ubiquitin peptide on the protein substrate [74].

5.1. Aurora-A and CUL3-KLHL18

The multiprotein complex E3 ubiquitin ligases of the cullin-RING-type ubiquitin ligase family include eight members in human. In the case of Cul3, the broad-complex, tramtrack and bric-a-brac (BTB) domain-containing proteins like KLHL18 (*Kelch-like*) serve to recognize the ubiquitin substrate [75]. There are about 200 BTB proteins in human with various functions not all being Cul3 adaptors [76]. Cul3 and KLHL18 localize at the centrosome in late G2, and the depletion of each of the protein provokes a delay in the G2/M transition that has been attributed to a default in Aurora-A phosphorylation on T288 and consequently a default in the kinase activation at the centrosome [77]. The activation of Aurora-A by Cul3-KLHL18 involved a nonproteasomal ubiquitination of the kinase; however, the activation is not a direct effect of ubiquitination (**Figure 2**). Although the mechanism of activation is not fully understood, the hypothesis is that ubiquitination of Aurora-A could facilitate the interaction of the kinase with its activators in late G2, such as Ajuba or Cep192 for instance (**Figure 1**).

5.2. Aurora-B and CUL3-BTB proteins

Aurora-B unlike Aurora-A binds to three different BTB proteins KLHL9, KLHL13 and KLHL21, and all three substrate adaptors participate to Cul3 complexes that ubiquitinate Aurora-B *in vivo* and *in vitro* [78, 79]. Like for Aurora-A, the ubiquitination of Aurora-B does not lead to any degradation of the kinase, although depletion of Cul3, KLHL9 and KLHL13 mimics depletion of the 26S proteasome [78]. On the contrary, ubiquitination by Cul3-KLHL9, −KLHL13 or KLHL21 regulates the kinase localization during mitosis. In the absence of KLHL9 or KLHL13, instead of moving to the kinetochore region, Aurora-B remains on chromosome arms during

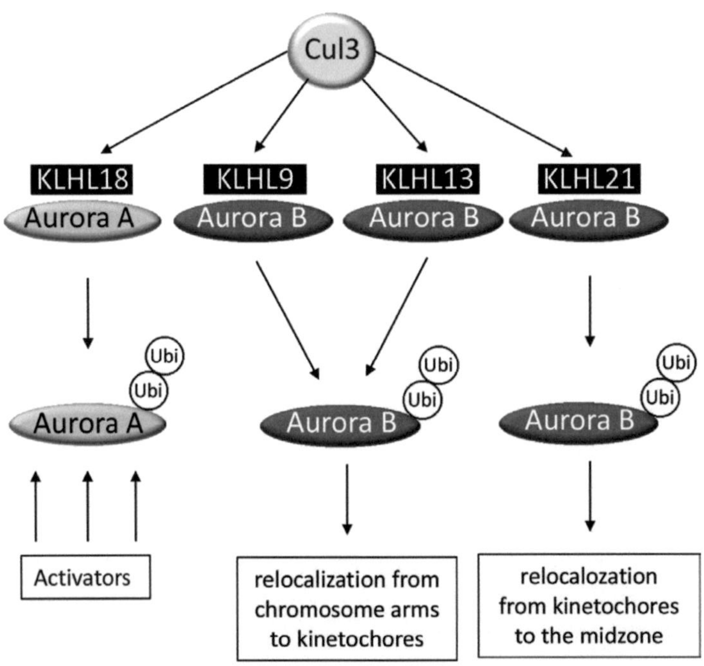

Figure 2. Cul3-dependent ubiquitination of Aurora kinases. Aurora-A is targeted by KLHL18, while Aurora-B by KLHL9, KLHL13 and KLHL21. Ubiquitination of Aurora-B localizes the protein, while ubiquitination of Aurora-A stimulates binding of the activators.

prometaphase/metaphase [78]. In the absence of KLHL21 instead of moving to the midzone, Aurora-B remains on anaphase chromosomes (**Figure 2**). In this last case, it is the whole CPC complex with Aurora-B, INCENP, Survivin and Borealin that remains on anaphase chromosomes [79].

Although KLHL9, KLHL13 and KLHL21 are substrate adaptors, their localizations do not really fit to their function. KLHL9 and KLHL13 have not been found on chromosomes for instance. On the other hand, KLHL21 does localize to the midzone where it could bring Cul3 to ubiquitinate Aurora-B, but even in this case KLHL21 should be at the kinetochore to ubiquitinate Aurora-B to target it to the midzone. The proposed hypothesis to explain this discrepancy is the high turnover of Aurora-B on its localization.

5.3. Aurora-A, Aurora-B and FBXW7

FBXW7 is an F-box protein participating to Skp, Cul1, F-box containing complex (SCF), a multiprotein E3 ubiquitin ligase complex that ubiquitinates proteins to be degraded by the proteasome. There are 69 F-box proteins coded by the human genome. Like KLHL proteins for Cul3, F-box proteins target the ubiquitin ligase complex to Cul1 substrates. FBXW7 binds to Aurora-A and Aurora-B and participates to their ubiquitination *in vivo* and *in vitro* [80, 81]. Depletions of FBXW7 stabilize both protein kinases levels *in vivo*, indicating that the FBXW7-dependent ubiquitination leads to the degradation of the kinases by the proteasome [81, 82]. F-box proteins usually recognize phosphorylated proteins, and FBXW7 for instance binds to Aurora-A previously phosphorylated by GSK3β on S245 and S387 [81] (**Figure 3**). Phosphorylation sites involved in Aurora-B/FBXW7 have not been identified, and the kinase involved is not known.

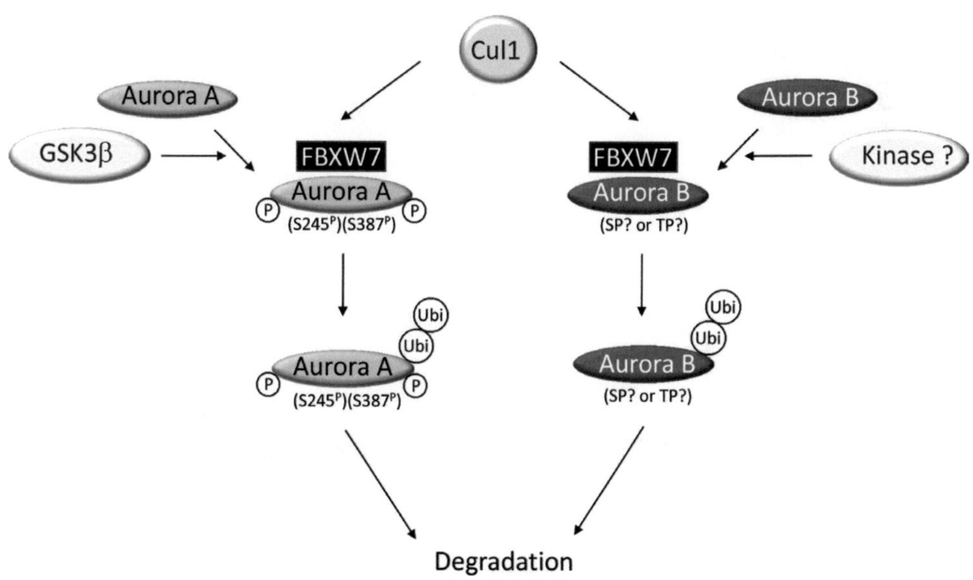

Figure 3. Cul1-dependent ubiquitination of Aurora kinases. Both Aurora-A and Aurora-B are targeted by FBXW7 F-box for ubiquitination. Aurora-A must be previously phosphorylated by GSK3β, while the kinase phosphorylating Aurora-B is unknown. Ubiquitination triggers degradation.

5.4. Aurora-A, Aurora-B and Cdh1

Cdc20-homologue 1 (Cdh1) is an activator and substrate adaptor of the E3 Ubiquitin ligase APC/C (Anaphase Promoting Complex/cyclostome). Cdh1 recognizes proteins containing a D-box (destruction box) (motif…RxxL…) and a KEN-box (motif …KEN…). All three Aurora kinases (A, B and C) contain a D-box in the carboxy end of the protein [83]. Both Aurora-A and B are ubiquitinated and degraded in a D-box-dependent manner [83, 84] although there are conflicting reports regarding Aurora-B [85]. Whether Aurora-C is degraded through its D-box remains also an open question. The Cdh1-dependent degradation of Aurora-A was demonstrated in both *Xenopus* and *human* [86, 87]. A new sequence required for the Cdh1-dependent degradation was discovered in Aurora-A from different species that was absent in Aurora-B and C. This sequence in the NH2 terminal end of the protein was named simultaneously A-box (for only in Aurora-A) [88] and DAD-box (for D-box-activating domain) (**Figure 4**) [89]. Interestingly, the A-box of Aurora-A contains a Serine at position 51 that when phosphorylated stabilizes the protein by inhibiting the functionality of the A-box [86, 90].

When tested in *Xenopus laevis* extracts Aurora-B was not found to be degraded in a Cdh1-dependent manner, it does contain a D-box but no A-Box, and only a chimera protein Aurora-A/Aurora-B containing Aurora-A A-box and Aurora-B D-box could be degraded in the extract [88, 89]. However, study in human finally revealed that Aurora-B contains also a functional D-box recognized by Cdh1 [84]. Interestingly, the same authors report that the KEN-box in

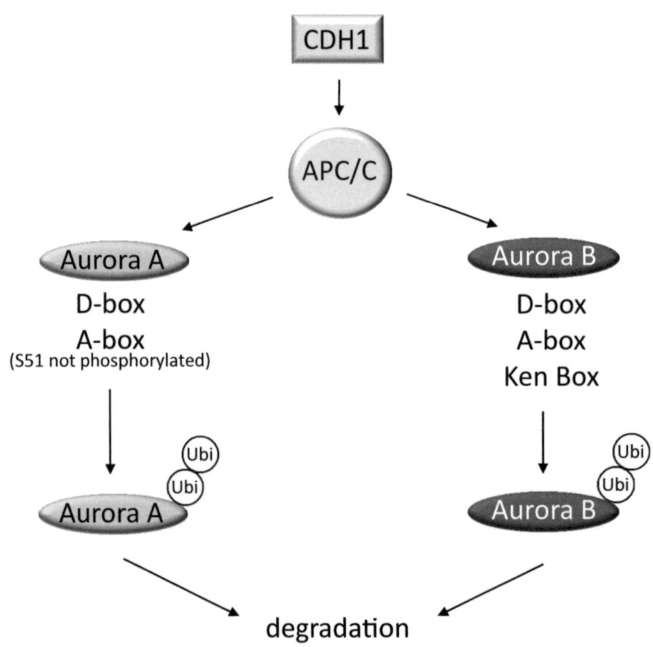

Figure 4. CDH1-dependent ubiquitination of Aurora kinases by APC/C. Both Aurora-A and Aurora-B are targeted by APC/C to be degraded. The ubiquitination required the presence A- and D-boxes in Aurora-A as well as the unphosphorylated state of S51 and the presence A- and D- and KEN-boxes in Aurora-B.

Aurora-B is required for Cdh1-dependent degradation. The discovery of a functional A-box in Aurora-B was more surprising, and since the box is not only present in Aurora-A, the name DAD-box seems now more adequate than A-box to name it. Finally, Aurora-B needs three functional boxes to be degraded in a Cdh1-dependent manner from the NH2 to the COOH end: a KEN-Box (KEN), an A-box (QRVL) and a D-box (RxxL) [85].

6. Sumoylation

Sumoylation resembles ubiquitination, and it is a posttranslational modification corresponding to a covalent attachment of a one or several SUMO proteins (100 amino acids) to a substrate. SUMOs stands for small ubiquitin-related modifiers, and there are now about ten different ubiquitin-like modifiers including ubiquitin and SUMO [91]. Like for ubiquitination, the sites of sumoylation are lysine residues, and the modification occurs also in a three-step reaction by E1, E2 and E3 ligases.

Both Aurora-A and Aurora-B are sumoylated *in vivo* on a lysine residue located in the sequence ...IHDRIKPEN... conserved in all Aurora kinases [92, 93].

Aurora-A is sumoylated on K249, and expression of the Aurora-A mutant K249R that cannot be sumoylated affects spindle assembly and potentiates the oncogenic property of the kinase [93]. Aurora-B is sumoylated on lysine 207, and expression of the Aurora-B mutant K207R affects chromosome segregation and cytokinesis [92]. Interestingly enough, sumoylation of Aurora-A or Aurora-B does not affect the kinase activity in vitro indicating that sumoylation is probably playing required for the localization of the protein or for protein-protein interaction. The exact function of Aurora sumoylation remains to be found.

7. Poly(ADP-ribosyl)ation

Poly(ADP-ribosyl)ation (PARylation) is a covalent modification of protein that can be catalyzed by 16 different Poly(ADP-ribose) polymerases (PARP) that attach multiple ADP-ribose units on substrate proteins by hydrolyzing NAD+. The reverse reaction is insured by Poly(ADP-ribose) glycohydrolase (PARG) (three isoforms) and several ADP ribose hydrolases [94, 95]. Only Aurora-B is PARylated *in vivo*, and interestingly enough, the modification occurs in the presence of DNA damage [96]. PARylation of Aurora-B leads to a decrease of Serine 10 histone H3 phosphorylation, indicating a loss of Aurora-B kinase activity. The kinase directly interacts with PARP-1 and PARP-2, and both enzymes can PARylate Aurora-B and inhibit its activity [96].

8. Phosphorylation

Aurora kinases belong to a family of protein kinases that need to be phosphorylated on a threonine residue in its T-loop to be active [97]. Aurora-A must be phosphorylated on T288,

Aurora-B on T232 and Aurora-C on T198 and T202 [37, 98, 99]. These phosphorylations are autophosphorylation events that occur in the presence of an activator such as TPX2 for Aurora-A or INCENP for Aurora-B and C [16, 99]. Aurora-A has two other levels of regulation controlled by phosphorylation. The kinase activity can be upregulated by autophosphorylation of S89 in the presence of nucleophosmin, as described above [50]. Aurora-A can also be downregulated by phosphorylation of S342 [64]. *In vivo*, phosphorylation of S342 occurs in the presence of DNA damages during G2 and depends on the activation of the checkpoint kinase Chk1 [66, 100]. Is Aurora-A a direct target of Chk1 or the target of a kinase downstream of Chk1 remains an open question.

Since Aurora kinases are phosphorylated, they are obviously targeted by phosphatases. T288 in Aurora-A and T232 in Aurora-B for instance are dephosphorylated and inactivated by type 1 protein phosphatase [101, 102]. T288 is also dephosphorylated by PPP6 that specifically target Aurora-A when bound to TPX2 [103]. These dephosphorylations inactivated Aurora kinase activity.

As discussed above, Aurora-A degradation by APC/C-CDH1 depends on the presence of an A-box in the kinase and phosphorylation of S51 within the human A-box inhibits this degradation process [64]. Although the kinase responsible for this phosphorylation remains to be identified, the phosphatase PP2A insures its dephosphorylation [104].

9. Conclusion

Since their discovery, Aurora kinases have become priority targets for the development of inhibitors for cancer treatments [105]. But their regulation takes multiple forms, adding difficulties in developing the efficient drugs targeting the kinases. This review tends to report a nonexhaustive list of posttranslational modifications (PTMs) affecting the functions of the kinases. These PTMs can be used as biomarkers, like the phosphorylation of T288 in Aurora-A frequently used to measure the kinase activity *in vivo*, and this test is currently questioned [106]. More interestingly, these PTMs can be used to design original inhibitory strategies different from those targeting the kinase active site. The binding of TPX2 to Aurora-A for instance has been targeted to search for Aurora-A inhibitor [107]. This kind of approach targeting PTMs offers broad prospects for specific inhibition of Aurora kinases. Many new inhibitors should then be discovered in the coming years.

Acknowledgements

We wish to apologize to the authors who were not quoted in this review. Work in the laboratory is supported by grants from the "Ligue Nationale Contre le Cancer" (LNCC, équipe labelisée 2014–2016 and 2017), the CNRS and the University of Rennes 1. APD venus was supported by "La Fondation Rennes 1." LV, APD are fellow of the LNCC and Région Bretagne, while OG was supported by "La Fondation pour la Recherche Médicale" (FRM).

Author details

Lucie Vaufrey, Arun Prasath Damodaran, Olivia Gavard, Stéphanie Le Bras and Claude Prigent*

*Address all correspondence to: claude.prigent@univ-rennes1.fr

University of Rennes 1, CNRS, Institute of Genetics and Development of Rennes, Equipe Labelisée Ligue Nationale Contre le Cancer, Rennes, France

References

[1] Ma HT, Poon RY. How protein kinases co-ordinate mitosis in animal cells. Biochemical Journal. 2011;**435**:17-31

[2] Salaun P, Rannou Y, Prigent C. Cdk1, Plks, auroras, and neks: The mitotic bodyguards. Advances in Experimental Medicine and Biology. 2008;**617**:41-56

[3] Goldenson B, Crispino JD. The Aurora kinases in cell cycle and leukemia. Oncogene. 2015;**34**:537-545

[4] Prosser SL, Pelletier L. Mitotic spindle assembly in animal cells: A fine balancing act. Nature Reviews Molecular Cell Biology. 2017;**18**:187-201

[5] Schiel JA, Prekeris R. Making the final cut — Mechanisms mediating the abscission step of cytokinesis. Scientific World Journal. 2010;**10**:1424-1434

[6] Reboutier D, Benaud C, Prigent C. Aurora A's functions during mitotic exit: The guess who game. Frontiers in Oncology. 2015;**5**:290

[7] Ruchaud S, Carmena M, Earnshaw WC. Chromosomal passengers: Conducting cell division. Nature Reviews Molecular Cell Biology. 2007;**8**(10):798-812

[8] Krenn V, Musacchio A. The Aurora B kinase in chromosome bi-orientation and spindle checkpoint signaling. Frontiers in Oncology. 2015;**5**:225

[9] Yang KT, Tang CJ, Tang TK. Possible role of Aurora-C in meiosis. Frontiers in Oncology. 2015;**5**:178

[10] Heidebrecht HJ, Buck F, Steinmann J, Sprenger R, Wacker HH, Parwaresch R. p100: A novel proliferation-associated nuclear protein specifically restricted to cell cycle phases S, G2, and M. Blood. 1997;**90**:226-233

[11] Gruss OJ, Wittmann M, Yokoyama H, Pepperkok R, Kufer T, Silljé H, Karsenti E, Mattaj IW, Vernos I. Chromosome-induced microtubule assembly mediated by TPX2 is required for spindle formation in HeLa cells. Nature Cell Biology. 2002;**4**:871-879

[12] Wittmann T, Boleti H, Antony C, Karsenti E, Vernos I. Localization of the kinesin-like protein Xklp2 to spindle poles requires a leucine zipper, a microtubule-associated protein, and dynein. Journal of Cell Biology. 1998;**143**:673-685

[13] Boleti H, Karsenti E, Vernos I. Xklp2, a novel Xenopus centrosomal kinesin-like protein required for centrosome separation during mitosis. Cell. 1996;**84**:49-59

[14] Wittmann T, Wilm M, Karsenti E, Vernos I. TPX2, A novel xenopus MAP involved in spindle pole organization. Journal of Cell Biology. 2000;**149**:1405-1418

[15] Kufer TA, Silljé HH, Körner R, Gruss OJ, Meraldi P, Nigg EA. Human TPX2 is required for targeting Aurora-A kinase to the spindle. Journal of Cell Biology. 2002;**158**:617-623

[16] Tsai MY, Wiese C, Cao K, Martin O, Donovan P, Ruderman J, Prigent C, Zheng Y. A Ran signalling pathway mediated by the mitotic kinase Aurora A in spindle assembly. Nature Cell Biology. 2003;**5**:242-248

[17] Eyers PA, Erikson E, Chen LG, Maller JL. A novel mechanism for activation of the protein kinase Aurora A. Current Biology. 2003;**13**:691-697

[18] Bayliss R, Sardon T, Vernos I, Conti E. Structural basis of Aurora-A activation by TPX2 at the mitotic spindle. Molecular Cell. 2003;**12**:851-862

[19] Dodson CA, Bayliss R. Activation of Aurora-A kinase by protein partner binding and phosphorylation are independent and synergistic. The Journal of Biological Chemistry. 2002;**287**:1150-1157

[20] Anderson K, Yang J, Koretke K, Nurse K, Calamari A, Kirkpatrick RB, Patrick D, Silva D, Tummino PJ, Copeland RA, Lai Z. Binding of TPX2 to Aurora A alters substrate and inhibitor interactions. Biochemistry. 2007;**46**:10287-10295

[21] Andersen JS, Wilkinson CJ, Mayor T, Mortensen P, Nigg EA, Mann M. Proteomic characterization of the human centrosome by protein correlation profiling. Nature. 2003;**426**:570-574

[22] Gomez-Ferreria MA, Rath U, Buster DW, Chanda SK, Caldwell JS, Rines DR, Sharp DJ. Human Cep192 is required for mitotic centrosome and spindle assembly. Current Biology. 2007;**17**:1960-1966

[23] Roghi C, Giet R, Uzbekov R, Morin N, Chartrain I, Le Guellec R, Couturier A, Dorée M, Philippe M, Prigent C. The Xenopus protein kinase pEg2 associates with the centrosome in a cell cycle-dependent manner, binds to the spindle microtubules and is involved in bipolar mitotic spindle assembly. Journal of Cell Science. 1998;**111**:557-572

[24] Schumacher JM, Ashcroft N, Donovan PJ, Golden AA. highly conserved centrosomal kinase, AIR-1, is required for accurate cell cycle progression and segregation of developmental factors in *Caenorhabditis elegans* embryos. Development. 1998;**125**:4391-4402

[25] Joukov V, De Nicolo A, Rodriguez A, Walter JC, Livingston DM. Centrosomal protein of 192 kDa (Cep192) promotes centrosome-driven spindle assembly by engaging in

organelle-specific Aurora A activation. Proceedings of the National Academy of Science United States of America. 2010;**107**:21022-21027

[26] Stebbins-Boaz B, Cao Q, de Moor CH, Mendez R, Richter JD. Maskin is a CPEB-associated factor that transiently interacts with elF-4E. Molecular Cell. 1999;**4**:1017-1027

[27] Pascreau G, Delcros JG, Cremet JY, Prigent C, Arlot-Bonnemains Y. Phosphorylation of maskin by Aurora-A participates in the control of sequential protein synthesis during *Xenopus laevis* oocyte maturation. The Journal of Biological Chemistry. 2005;**280**:13415-13423

[28] Giet R, McLean D, Descamps S, Lee MJ, Raff JW, Prigent C, Glover DM. Drosophila Aurora A kinase is required to localize D-TACC to centrosomes and to regulate astral microtubules. Journal of Cell Biology. 2002;**156**:437-451

[29] Kinoshita K, Noetzel TL, Pelletier L, Mechtler K, Drechsel DN, Schwager A, Lee M, Raff JW, Hyman AA. Aurora A phosphorylation of TACC3/maskin is required for centrosome-dependent microtubule assembly in mitosis. Journal of Cell Biology. 2005;**170**:1047-1055

[30] Barros TP, Kinoshita K, Hyman AA, Raff JW. Aurora A activates D-TACC-Msps complexes exclusively at centrosomes to stabilize centrosomal microtubules. Journal of Cell Biology. 2005;**170**:1039-1046

[31] Peset I, Seiler J, Sardon T, Bejarano LA, Rybina S, Vernos I. Function and regulation of Maskin, a TACC family protein, in microtubule growth during mitosis. Journal of Cell Biology. 2005;**170**:1057-1066

[32] Lioutas A, Vernos I. Aurora A kinase and its substrate TACC3 are required for central spindle assembly. EMBO Reports. 2013;**14**:829-836

[33] Carmena M, Wheelock M, Funabiki H, Earnshaw WC. The chromosomal passenger complex (CPC): From easy rider to the godfather of mitosis. Nature Reviews Molecular Cell Biology. 2012;**13**:789-803

[34] Trivedi P, Stukenberg PT. A centromere-signaling network underlies the coordination among mitotic events. Trends in Biochemical Sciences. 2016;**41**:160-174

[35] Adams RR, Wheatley SP, Gouldsworthy AM, Kandels-Lewis SE, Carmena M, Smythe C, Gerloff DL, Earnshaw WC. INCENP binds the Aurora-related kinase AIRK2 and is required to target it to chromosomes, the central spindle and cleavage furrow. Current Biology. 2000;**10**:1075-1078

[36] Kaitna S, Mendoza M, Jantsch-Plunger V, Glotzer M. Incenp and an aurora-like kinase form a complex essential for chromosome segregation and efficient completion of cytokinesis. Current Biology. 2000;**10**:1172-1181

[37] Yasui Y, Urano T, Kawajiri A, Nagata K, Tatsuka M, Saya H, Furukawa K, Takahashi T, Izawa I, Inagaki M. Autophosphorylation of a newly identified site of Aurora-B is indispensable for cytokinesis. The Journal of Biological Chemistry. 2004;**279**:12997-13003

[38] Yang J, Zappacosta F, Annan RS, Nurse K, Tummino PJ, Copeland RA, Lai Z. The catalytic role of INCENP in Aurora B activation and the kinetic mechanism of Aurora B/INCENP. Biochemical Journal. 2009;**417**:355-360

[39] Eyers PA, Churchill ME, Maller JL. The Aurora A and Aurora B protein kinases: A single amino acid difference controls intrinsic activity and activation by TPX2. Cell Cycle. 2005;**4**:784-789

[40] Hans F, Skoufias DA, Dimitrov S, Margolis RL. Molecular distinctions between Aurora A and B: A single residue change transforms Aurora A into correctly localized and functional Aurora B. Molecular Biology of the Cell. 2009;**20**:3491-3502

[41] Goshima G. Identification of a TPX2-like microtubule-associated protein in Drosophila. PLoS One. 2011;**6**:e28120

[42] Bach I. The LIM domain: Regulation by association. Mechanisms of Development. 2000; **91**:5-17

[43] Hirota T, Kunitoku N, Sasayama T, Marumoto T, Zhang D, Nitta M, Hatakeyama K, Saya H. Aurora-A and an interacting activator, the LIM protein Ajuba, are required for mitotic commitment in human cells. Cell. 2003;**114**:585-598

[44] Bai M, Ni J, Wu J, Wang B, Shen S, Yu L. A novel mechanism for activation of Aurora-A kinase by Ajuba. Gene. 2014;**543**:133-139

[45] Eckerdt F, Pascreau G, Phistry M, Lewellyn AL, DePaoli-Roach AA, Maller JL. Phosphorylation of TPX2 by Plx1 enhances activation of Aurora A. Cell Cycle. 2009; **8**:2413-2419

[46] Sabino D, Brown NH, Basto R. Drosophila Ajuba is not an Aurora-A activator but is required to maintain Aurora-A at the centrosome. Journal of Cell Science. 2011;**124**: 1156-1166

[47] Box JK, Paquet N, Adams MN, Boucher D, Bolderson E, O'Byrne KJ, Richard DJ. Nucleophosmin: From structure and function to disease development. BMC Molecular Biology. 2016;**17**:19

[48] Okuda M, Horn HF, Tarapore P, Tokuyama Y, Smulian AG, Chan PK, Knudsen ES, Hofmann IA, Snyder JD, Bove KE, Fukasawa K. Nucleophosmin/B23 is a target of CDK2/cyclin E in centrosome duplication. Cell. 2000;**103**:127-140

[49] Amin MA, Matsunaga S, Uchiyama S, Fukui K. Nucleophosmin is required for chromosome congression, proper mitotic spindle formation, and kinetochore-microtubule attachment in HeLa cells. FEBS Letters. 2008;**582**:3839-3844

[50] Reboutier D, Troadec MB, Cremet JY, Fukasawa K, Prigent C. Nucleophosmin/B23 activates Aurora A at the centrosome through phosphorylation of serine 89. Journal of Cell Biology. 2012;**197**:19-26

[51] Hutterer A, Berdnik D, Wirtz-Peitz F, Zigman M, Schleiffer A, Knoblich JA. Mitotic activation of the kinase Aurora-A requires its binding partner Bora. Developmental Cell. 2006;**11**:147-157

[52] Seki A, Coppinger JA, Jang CY, Yates JR, Fang G. Bora and the kinase Aurora a cooperatively activate the kinase Plk1 and control mitotic entry. Science. 2008;**320**:1655-1658

[53] Macůrek L, Lindqvist A, Lim D, Lampson MA, Klompmaker R, Freire R, Clouin C, Taylor SS, Yaffe MB, Medema RH. Polo-like kinase-1 is activated by Aurora A to promote checkpoint recovery. Nature. 2008;**455**:119-123

[54] Lieu AS, Cheng TS, Chou CH, CH W, Hs CY, Huang CY, Chang LK, Loh JK, Chang CS, Hsu CM, Howng SL, Hong YR. Functional characterization of AIBp, a novel Aurora-A binding protein in centrosome structure and spindle formation. International Journal of Oncology. 2010;**37**:429-436

[55] Chou CH, Loh JK, Yang MC, Lin CC, Hong MC, Cho CL, Chou AK, Wang CH, Lieu AS, Howng SL, Hsu CM, Hong YR. AIBp regulates mitotic entry and mitotic spindle assembly by controlling activation of both Aurora-A and Plk1. Cell Cycle. 2015;**14**:2764-2776

[56] Singh M, Cowell L, Seo S, O'Neill G, Golemis E. Molecular basis for HEF1/NEDD9/Cas-L action as a multifunctional co-ordinator of invasion, apoptosis and cell cycle. Cell Biochemistry and Biophysics. 2007;**48**:54-72

[57] Pugacheva EN, Golemis EA. The focal adhesion scaffolding protein HEF1 regulates activation of the Aurora-A and Nek2 kinases at the centrosome. Nature Cell Biology. 2005;**7**:937-946

[58] Pugacheva EN, Jablonski SA, Hartman TR, Henske EP, Golemis EA. HEF1-dependent Aurora A activation induces disassembly of the primary cilium. Cell. 2007;**129**:1351-1363

[59] Plotnikova OV, Nikonova AS, Loskutov YV, Kozyulina PY, Pugacheva EN, Golemis EA. Calmodulin activation of Aurora-A kinase (AURKA) is required during ciliary disassembly and in mitosis. Molecular Biology of the Cell. 2012;**23**:2658-2670

[60] Manser E, Leung T, Salihuddin H, Zhao ZS, Lim L. A brain serine/threonine protein kinase activated by Cdc42 and Rac1. Nature. 1994;**367**:40-46

[61] Sells MA, Boyd JT, Chernoff J. p21-Activated kinase 1 (Pak1) regulates cell motility in mammalian fibroblasts. Journal of Cell Biology. 1999;**145**(4):837-849

[62] Manser E, Huang HY, Loo TH, Chen XQ, Dong JM, Leung T, Lim L. Expression of constitutively active alpha-PAK reveals effects of the kinase on actin and focal complexes. Molecular and Cell Biology. 1997;**17**:1129-1143

[63] Zhao ZS, Lim JP, Ng YW, Lim L, Manser E. The GIT-associated kinase PAK targets to the centrosome and regulates Aurora-A. Molecular Cell. 2005;**20**:237-249

[64] Littlepage LE, Ruderman JV. Identification of a new APC/C recognition domain, the A box, which is required for the Cdh1-dependent destruction of the kinase Aurora-A during mitotic exit. Genes & Development. 2002;**16**:2274-2285

[65] Pascreau G, Delcros JG, Morin N, Prigent C, Arlot-Bonnemains Y. Aurora-A kinase Ser349 phosphorylation is required during *Xenopus laevis* oocyte maturation. Developmental Biology. 2008;**317**:523-530

[66] Krystyniak A, Garcia-Echeverria C, Prigent C, Ferrari S. Inhibition of Aurora A in response to DNA damage. Oncogene. 2006;**25**:338-348

[67] Hannigan GE, Leung-Hagesteijn C, Fitz-Gibbon L, Coppolino MG, Radeva G, Filmus J, Bell JC, Dedhar S. Regulation of cell adhesion and anchorage-dependent growth by a new beta 1-integrin-linked protein kinase. Nature. 1996;**379**:91-96

[68] Wu C, Dedhar S. Integrin-linked kinase (ILK) and its interactors, A new paradigm for the coupling of extracellular matrix to actin cytoskeleton and signaling complexes. Journal of Cell Biology. 2001;**155**:505-510

[69] Fielding AB, Dobreva I, McDonald PC, Foster LJ, Dedhar S. Integrin-linked kinase localizes to the centrosome and regulates mitotic spindle organization. Journal of Cell Biology. 2008;**180**:681-689

[70] Fielding AB, Lim S, Montgomery K, Dobreva I, Dedhar S. A critical role of integrin-linked kinase, ch-TOG and TACC3 in centrosome clustering in cancer cells. Oncogene. 2011;**30**:521-534

[71] Burgess SG, Peset I, Joseph N, Cavazza T, Vernos I, Pfuhl M, Gergely F, Bayliss R. Aurora-A-dependent control of TACC3 influences the rate of mitotic spindle assembly. PLoS Genetics. 2015;**11**:e1005345

[72] Pizarro-Cerdá J, Chorev DS, Geiger B, Cossart P. The diverse family of Arp2/3 complexes. Trends in Cell Biology. 2017;**27**:93-100

[73] Molli PR, Li DQ, Bagheri-Yarmand R, Pakala SB, Katayama H, Sen S, Iyer J, Chernoff J, Tsai MY, Nair SS, Kumar R. Arpc1b, a centrosomal protein, is both an activator and substrate of Aurora A. Journal of Cell Biology. 2010;**190**:101-114

[74] Swatek KN, Komander D. Ubiquitin modifications. Cell Research. 2016;**26**:399-422

[75] Fujiyama-Nakamura S, Ito S, Sawatsubashi S, Yamauchi Y, Suzuki E, Tanabe M, Kimura S, Murata T, Isobe T, Takeyama K, Kato S. BTB protein, dKLHL18/CG3571,serves as an adaptor subunit for a dCul3 ubiquitin ligase complex. Genes to Cells. 2009;**14**:965-973

[76] Chen HY, Chen RH. Cullin 3 ubiquitin ligases in cancer biology: Functions and therapeutic implications. Frontiers in Oncology. 2016;**6**:113

[77] Moghe S, Jiang F, Miura Y, Cerny RL, Tsai MY, Furukawa M. The CUL3-KLHL18 ligase regulates mitotic entry and ubiquitylates Aurora-A. Biology Open. 2012;**1**:82-91

[78] Sumara I, Quadroni M, Frei C, Olma MH, Sumara G, Ricci R, Peter M. A Cul3-based E3 ligase removes Aurora B from mitotic chromosomes, regulating mitotic progression and completion of cytokinesis in human cells. Developmental Cell. 2007;**12**:887-900

[79] Maerki S, Olma MH, Staubli T, Steigemann P, Gerlich DW, Quadroni M, Sumara I, Peter M. The Cul3-KLHL21 E3 ubiquitin ligase targets Aurora B to midzone microtubules in anaphase and is required for cytokinesis. Journal of Cell Biology. 2009;**187**:791-800

[80] Teng CL, Hsieh YC, Phan L, Shin J, Gully C, Velazquez-Torres G, Skerl S, Yeung SC, Hsu SL, Lee MH. FBXW7 is involved in Aurora B degradation. Cell Cycle. 2012;**11**:4059-4068

[81] Kwon YW, Kim IJ, Wu D, Lu J, Stock WA Jr, Liu Y, Huang Y, Kang HC, DelRosario R, Jen KY, Perez-Losada J, Wei G, Balmain A, Mao JH. Pten regulates Aurora-A and cooperates with Fbxw7 in modulating radiation-induced tumor development. Molecular Cancer Research. 2012;**10**:834-844

[82] Hagedorn M, Delugin M, Abraldes I, Allain N, Belaud-Rotureau MA, Turmo M, Prigent C, Loiseau H, Bikfalvi A, Javerzat S. FBXW7/hCDC4 controls glioma cell proliferation in vitro and is a prognostic marker for survival in glioblastoma patients. Cell Division. 2007;**2**:9

[83] Arlot-Bonnemains Y, Klotzbucher A, Giet R, Uzbekov R, Bihan R, Prigent C. Identification of a functional destruction box in the *Xenopus laevis* Aurora-A kinase pEg2. FEBS Letters. 2001;**508**:149-152

[84] Stewart S, Fang G. Destruction box-dependent degradation of Aurora B is mediated by the anaphase-promoting complex/cyclosome and Cdh1. Cancer Research. 2005;**65**:8730-8735

[85] Nguyen HG, Chinnappan D, Urano T, Ravid K. Mechanism of Aurora-B degradation and its dependency on intact KEN and A-boxes: Identification of an aneuploidy-promoting property. Molecular and Cell Biology. 2005;**25**:4977-4992

[86] Castro A, Arlot-Bonnemains Y, Vigneron S, Labbé JC, Prigent C, Lorca T. APC/fizzy-related targets Aurora-A kinase for proteolysis. EMBO Reports. 2002;**3**:457-462

[87] Si T, Honda K, Sugiura K, Yamaguchi A, Furukawa K, Urano T. Degradation of human Aurora-A protein kinase is mediated by hCdh1. FEBS Letters. 2002;**519**:59-65

[88] Littlepage LE, Wu H, Andresson T, Deanehan JK, Amundadottir LT, Ruderman JV. Identification of phosphorylated residues that affect the activity of the mitotic kinase Aurora-A. Proceedings of the National Academy of Science United States of America. 2002;**99**:15440-15445

[89] Castro A, Vigneron S, Bernis C, Labbé JC, Prigent C, Lorca T. The D-box-activating domain (DAD) is a new proteolysis signal that stimulates the silent D-box sequence of Aurora-A. EMBO Reports. 2002;**3**:1209-1214

[90] Crane R, Kloepfer A, Ruderman JV. Requirements for the destruction of human Aurora-A. Journal of Cell Science. 2004;**117**:5975-5983

[91] Eifler K, Vertegaal AC. SUMOylation-mediated regulation of cell cycle progression and cancer. Trends in Biochemical Sciences. 2015;**40**:779-793

[92] Fernández-Miranda G, Pérez de Castro I, Carmena M, Aguirre-Portolés C, Ruchaud S, Fant X, Montoya G, Earnshaw WC, Malumbres M. SUMOylation modulates the function of Aurora-B kinase. Journal of Cell Science. 2010;**123**:2823-2833

[93] Pérez de Castro I, Aguirre-Portolés C, Martin B, Fernández-Miranda G, Klotzbucher A, Kubbutat MH, Megías D, Arlot-Bonnemains Y, Malumbres M. A SUMOylation motif in Aurora-A: Implications for spindle dynamics and oncogenesis. Frontiers in Oncology. 2011;**1**:50

[94] Gibson BA, Kraus WL. New insights into the molecular and cellular functions of poly(ADP-ribose) and PARPs. Nature Reviews Molecular Cell Biology. 2012;**13**:411-424

[95] Li N, Chen J. ADP-ribosylation: Activation, recognition, and removal. Molecules and Cells. 2014;**37**:9-16

[96] Monaco L, Kolthur-Seetharam U, Loury R, Murcia JM, de Murcia G, Sassone-Corsi P. Inhibition of Aurora-B kinase activity by poly(ADP-ribosyl)ation in response to DNA damage. Proceedings of the National Academy of Science United States of America. 2005;**102**:14244-14248

[97] Nolen B, Taylor S, Ghosh G. Regulation of protein kinases: controlling activity through activation segment conformation. Molecular Cell. 2004;**15**:661-675

[98] Walter AO, Seghezzi W, Korver W, Sheung J, Lees E. The mitotic serine/threonine kinase Aurora2/AIK is regulated by phosphorylation and degradation. Oncogene. 2000;**19**:4906-4916

[99] Sasai K, Katayama H, Hawke DH, Sen S. Aurora-C interactions with survivin and INCENP reveal shared and distinct features compared with Aurora-B chromosome passenger protein complex. PLoS One. 2016;**11**:e0157305

[100] Cazales M, Schmitt E, Montembault E, Dozier C, Prigent C, Ducommun B. CDC25B phosphorylation by Aurora-A occurs at the G2/M transition and is inhibited by DNA damage. Cell Cycle. 2005;**4**:1233-1238

[101] Katayama H, Zhou H, Li Q, Tatsuka M, Sen S. Interaction and feedback regulation between STK15/BTAK/Aurora-A kinase and protein phosphatase 1 through mitotic cell division cycle. The Journal of Biological Chemistry. 2001;**276**:46219-46224

[102] Murnion ME, Adams RR, Callister DM, Allis CD, Earnshaw WC, Swedlow JR. Chromatin-associated protein phosphatase 1 regulates Aurora-B and histone H3 phosphorylation. The Journal of Biological Chemistry. 2001;**276**:26656-26665

[103] Zeng K, Bastos RN, Barr FA, Gruneberg U. Protein phosphatase 6 regulates mitotic spindle formation by controlling the T-loop phosphorylation state of Aurora A bound to its activator TPX2. Journal of Cell Biology. 2010;**191**:1315-1332

[104] Horn V, Thélu J, Garcia A, Albigès-Rizo C, Block MR, Viallet J. Functional interaction of Aurora-A and PP2A during mitosis. Mol Biol Cell. 2007;**18**:1233-41

[105] Tang A, Gao K, Chu L, Zhang R, Yang J, Zheng J. Aurora kinases: Novel therapy targets in cancers. Oncotarget. 2017;**8**:23937-23954

[106] Shagisultanova E, Dunbrack RL Jr, Golemis EA. Issues in interpreting the in vivo activity of Aurora-A. Expert Opinion on Therapeutic Targets. 2015;**19**:187-200

[107] Gupta A, Jain R, Wahi D, Goyal S, Jamal S, Grover A. Abrogation of Aurora A-TPX2 by novel natural inhibitors: Molecular dynamics-based mechanistic analysis. Journal of Receptors and Signal Transduction Research. 2015;**35**:626-633

Section 2

Kinases and Diseases

Chapter 4

Biological Consequences of Priming Phosphorylation in Cancer Development

Katsuhiko Aoki and Kiyotsugu Yoshida

Additional information is available at the end of the chapter

http://dx.doi.org/10.5772/intechopen.70039

Abstract

Multisite phosphorylations on a single polypeptide mediated by protein kinase(s) are commonly observed. In some cases, hierarchical phosphorylations occur when first priming event triggers second processive phosphorylation. Hierarchal multisite phosphorylation that is mediated by a priming kinase and a processive kinase is a fail-safe system that accurately regulates physiological processes, including cell cycle progression, survival, migration, metabolism, differentiation and stem cell renewal. Here, we summarize the findings of cancer-associated priming kinases (CK1 and DYRK family) and processive kinase (GSK3). GSK3 has an unusual ability to accurately regulate the wide variety of cellular processes via the priming phosphorylation of its substrates. Therefore, dysregulation of priming phosphorylation gives rise to pathological disorders such as cancer.

Keywords: priming phosphorylation, multisite phosphorylation, hierarchical phosphorylation, priming kinase, CK1, DYRK1A, DYRK2, processive kinase, GSK3, NFAT signaling, Wnt signaling, β-catenin, SCF, β-TRCP, FBW7, LRP signalosome, protein stability, cancer

1. Introduction

Protein phosphorylation is the most frequent post-translational modification that regulates the function, interaction and stability of various proteins. Multisite phosphorylations on a single polypeptide, which are mediated by protein kinase(s), are commonly observed. In some cases, hierarchical phosphorylations occur when first phosphorylation event triggers second subsequent phosphorylation. Here, such a first phosphorylation is called as "priming phosphorylation." Priming phosphorylation is mediated by "priming kinase" and serial phosphorylation is mediated by "processive kinase" (**Figure 1**).

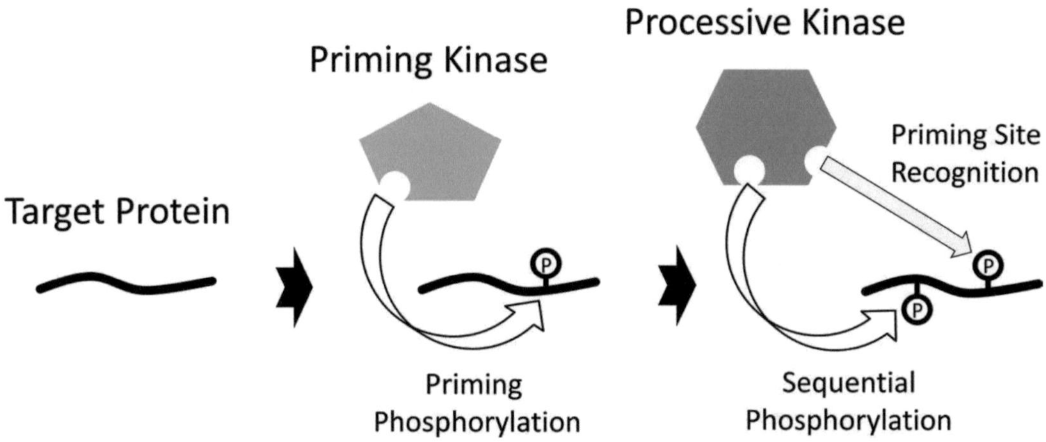

Figure 1. Schematic illustration of multisite phosphorylation mediated by a priming kinase and a processive kinase.

Processive kinases, such as glycogen synthase kinase 3 (GSK3), are ubiquitously expressed in mammalian tissues and involved in numerous cellular processes. In this context, priming kinases provide a basis for selective action of the individual cellular process regulated by processive kinases.

Hierarchical multisite phosphorylations by the priming kinase and the processive kinase are the fail-safe mechanism that accurately regulates physiological processes, including cell cycle progression, survival, migration, metabolism, differentiation and stem cell renewal. Therefore, loss of priming phosphorylation caused by impairment of priming kinases gives rise to pathological disorders, such as cancer.

2. Priming kinases

2.1. Casein kinase 1 (CK1)

The casein kinase 1 (CK1) family is evolutionary conserved serine/threonine protein kinases that are ubiquitously expressed in eukaryotic organisms from yeast to human [1]. In human, six CK1 isoforms (α, γ1, γ2, γ3, δ and ε) are encoded by distinct genes. These isoforms differ in length and sequence of N-terminal and C-terminal domain [2, 3].

The name casein kinase arose from the protein kinase activity using casein as an in vitro substrate [4]. Because casein is a highly phosphorylated protein, the casein kinase was initially characterized by a phosphate-directed protein kinase [5, 6]. However, it became evident that CK1 does not only phosphorylate phospho-primed substrates but also displays a prominent phosphorylation activity targeting the site that contains cluster of acidic amino acids, immediately N-terminal (−1 to −5) of the phospho-acceptor site [7–9]. The canonical consensus sequence for CK1 is shown in **Figure 2**.

http://dx.doi.org/10.5772/intechopen.70039

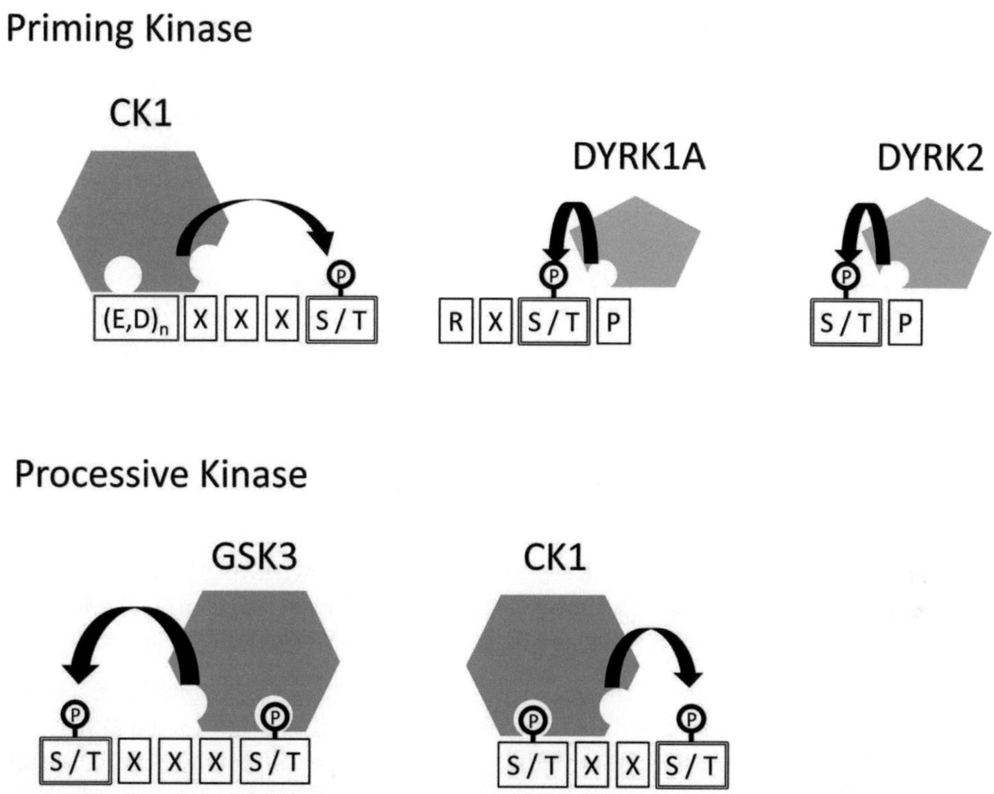

Figure 2. Consensus sequence of priming kinases and processive kinases. $(E,D)_n$ denotes acidic amino acids cluster; X denotes any amino acid and S/T denotes Ser or Thr. CK1 recognizes the negatively charged amino acid cluster. DYRK1A is a proline- and arginine-directed kinase. DYRK2 is a proline-directed kinase. GSK3 typically phosphorylates "primed" substrate that is pre-phosphorylated by a priming kinase. CK1 also behaves as a phosphate-directed processive kinase.

Members of the CK1 family are ubiquitously expressed but their expression levels differ in tissue and cell type [10–12]. Recently, an increasing number of substrate proteins have been identified, which are phosphorylated by CK1 family *in vitro* and *in vivo* [2, 13, 14]. According to a global weblogo analysis to a database of 35,000 non-redundant phosphosites, CK1 targets are responsible for the generation of 9.5% of the human phosphoproteome [15].

The wide range of substrates suggests that the members of CK1 family regulate diverse and important cellular functions. For instance, they are involved in Wnt signaling, Hedgehog signaling, Hippo signaling, neurodegenerative disease, circadian rhythms, vesicular trafficking, cytoskeleton dynamics, nuclear localization, DNA processing and repair, apoptosis, cell division, proliferation and differentiation [2, 13, 14, 16]. Consequently, deregulation or dysfunction of CK1 in these pathways responsible for growth, proliferation, and apoptosis may result in pathological condition, such as tumorigenesis [3, 17, 18]. CK1δ and CK1ε isoforms are overexpressed and activated in many tumor types, such as colon and pancreatic cancers [19, 20]. By contrast, the downregulation of CK1α leads to increased proliferation and invasive growth of melanoma cells [21, 22].

2.2. Dual specificity tyrosine phosphorylation-regulated kinases (DYRKs)

Dual specificity tyrosine phosphorylation-regulated kinase (DYRK) family is an evolutionally conserved eukaryotic protein kinases belong to CMGC protein kinase group [23, 24]. In human, five DYRK members (DYRK1A, DYRK1B, DYRK2, DYRK3 and DYRK4) are encoded by distinct genes [25].

Their kinase activity depends upon the autophosphorylation of a tyrosine residue in activation loop of catalytic domain [26, 27]. The autophosphorylation of tyrosine is an intramolecular event that is mediated by a short-lived translational intermediate of itself [28–30]. Once phosphorylated on the tyrosine residue, DYRKs lose tyrosine kinase ability and retain only serine/threonine kinase ability. Although DYRKs are potentially proline-directed kinase, they differ in their target recognition sequence and its preference for an arginine residue is a feature of DYRK1A but not of DYRK2 and DYRK4 [31–34]. The canonical consensus sequence for DYRK1A and DYRK2 is shown in **Figure 2**.

With some exceptions [35–37], DYRK1A and DYRK2 have been characterized as a potential tumor suppressor [38–49]. In contrast, DYRK1B (also known MIRK), closely related to DYRK1A, have been characterized as a positive regulator of cancer cell survival [50–61]. It is still not known to the details of the biological functions for DYRK3 and DYRK4.

Human DYRK1A is the most well-characterized member in the DYRK family [62, 63] because the gene is localized in the down syndrome (DS) critical region in chromosome 21 [64–67]. In mouse, DYRK1A is essential for embryonic development, especially in the nervous systems, and unbalanced gene dosage causes developmental delay and abnormal brain morphology [68–75]. In neuronal progenitor cells, overexpression of DYRK1A bring to the attenuation of cell proliferation that promotes the switch to a quiescent state or differentiation. DYRK1A mediates direct phosphorylation of p53 at Ser-15 that leads to the induction of p53 target genes, such as $p21^{CIP1}$, and impaired G1/G0-S phase transition [76].

It is known that individuals with DS have a significantly reduced incidence of solid tumors [77–79]. DS model mouse exhibits that considerable growth protection against transplantation of allogeneic tumors is caused by a deficit in tumor angiogenesis arising from suppression of nuclear factor of activated T cells (NFAT) transcriptional regulator pathway [38]. DYRK1A phosphorylates NFAT proteins in nucleus, thereby priming the subsequent phosphorylation by additional kinases (GSK3 and CK1), then fully phosphorylated NFAT proteins are exported from the nucleus to the cytoplasm (**Figure 3**). Cytoplasmic accumulation of NFAT proteins represses the NFAT pathway associated with tumor progression [80]. Paradoxically, children with DS have a remarkably increased risk of developing leukemias, including most types of acute megakaryoblastic leukemia (AMKL) and acute lymphoblastic leukemia (ALL). It has been suggested that DYRK1A is also a potent AMKL-promoting gene that modulates megakaryoblastic expansion through the inhibition of the NFAT pathway [35].

Although precise regulation of NFAT pathway is essential for vertebrate development and function, NFAT isoforms are overexpressed and activated in human solid tumors and leukemias.

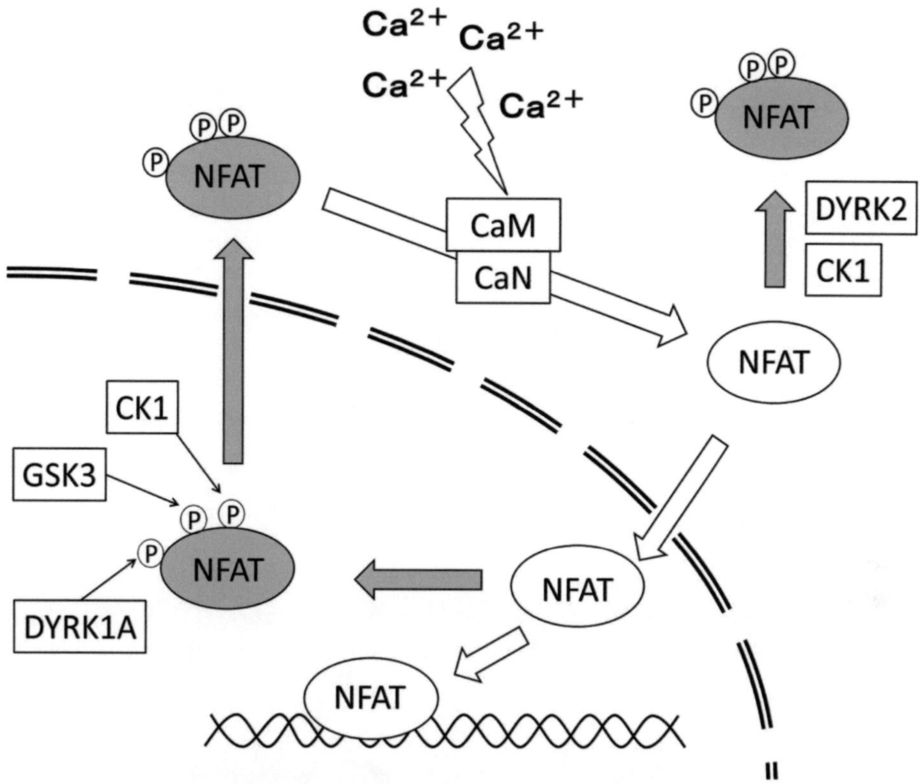

Figure 3. Schematic diagram of NFAT translocation. DYRK1A acts as an export kinase in the nucleus. DYRK2 acts as a maintain kinase in the cytoplasm. These kinases mediate the priming phosphorylation of NFAT proteins. Fully phosphorylated NFAT proteins are sequestered in cytoplasm. Increased intracellular Ca^{2+} levels activate the calmodulin (CaM)/calcineurin (CN) phosphatase complex. Dephosphorylated NFAT proteins relocate into nucleus and promote gene transcription.

This aberrant expression of NFAT proteins leads to the induction of the target genes that promote malignant phenotype that is associated with tumor progression, such as proliferation, survival, migration and invasion [80–82]. In the basal state, NFAT proteins are sequestered and inactivated as a phosphorylated form in the cytoplasm. DYRK1A acts as an export kinase that phosphorylates NFAT proteins inside the nucleus and induces its relocation to the cytoplasm. As a counterpart of export kinase, DYRK2 acts as a maintenance kinase that phosphorylates in the cytoplasm, where they keep NFAT proteins in a phosphorylated state, and prevents their translocation to the nucleus (**Figure 3**). DYRK1A and DYRK2 can directly phosphorylate the conserved SP3 motif of the NFAT regulatory domain and thereby can prime for the subsequent phosphorylation by GSK3 and/or CK1 [83].

Under normal conditions, DYRK2 is predominantly expressed in the cytoplasm and constitutively degraded by MDM2 ubiquitin ligase in the nucleus. Upon exposure to genotoxic stress, ATM protein kinase phosphorylates DYRK2 at Thr-33 and Ser-369. As a result, DYRK2 enable to escape from MDM2 and to induce the kinase activity toward p53 at Ser-46 in the nucleus. Phosphorylation of Ser-46 following severe DNA damage increases the transcriptional activity of pro-apoptotic genes [84, 85]. The other functional role of DYRK2 in DNA damage response may be link to DNA double-strand break repair pathway [86].

DYRK2 was found to be mutated in breast and central nervous system tumors, in nonsense and missense mutation, respectively [87, 88]. Loss of function of DYRK2 in cancer cells accelerated cell proliferation due to stabilization of oncogenic transcription factors, c-Jun and c-Myc [89]. This stabilization is caused by the loss of priming phosphorylation that is necessary to generate a phosphodegron that leads to subsequent SCF (Skp1-Cullin1-F-box protein) ubiquitin ligase-mediated degradation. Snail, a zinc finger protein to promote epithelial-mesenchymal transitions (EMT), is stabilized by the DYRK2 knockdown, probably in the same fashion as c-Jun/c-Myc, and allows cancer cells to represent loss of epithelial features and gain of invasiveness [90, 91] (**Figure 4**). Moreover, it is recently demonstrated that impairment of DYRK2 augments cancer stem-like traits of breast cancer cells [92].

2.3. Other priming kinases

It is also known that cAMP-dependent kinase (PKA), AMP-activated protein kinase (AMPK), cyclin-dependent kinase 5 (CDK5), DNA-dependent protein kinase (DNA-PK), calcium and calmodulin-dependent protein kinase II (CAMKII) and mitogen-activated protein kinases

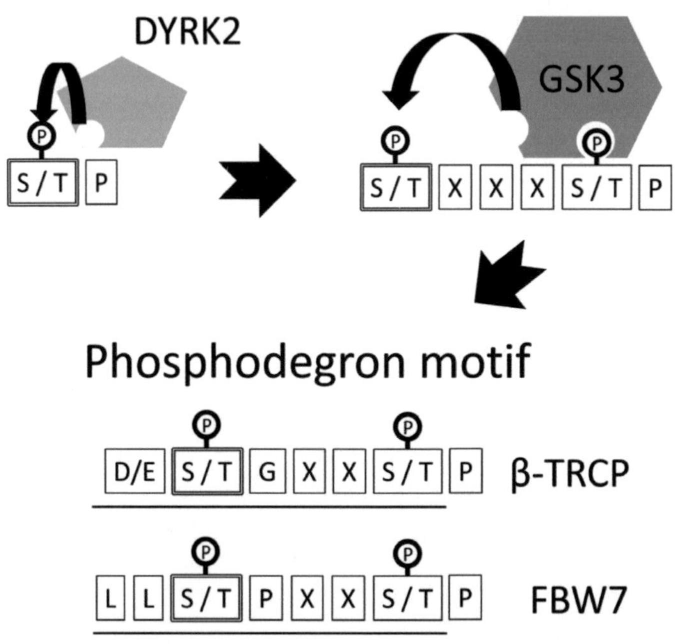

Figure 4. Consensus sequence of phosphodegron motif. These phosphodegron motifs are created by several priming kinases such as DYRK2 and processive kinase GSK3. Underlined sequence indicates the canonical sequence of β-TRCP and FBW7. Loss of priming phosphorylation leads to dysgenesis of phosphodegron and results in stabilization of their target proteins.

(MAPKs) can act as priming kinases for GSK3 [93]. CDK1 functions as a priming kinase for polo-like kinase 1 (PLK1) that is a key regulator of cell cycle progression [94, 95].

3. Processive kinases

Phosphate-directed protein kinases, such as CK1, CK2 [96, 97] and GSK3, function as the processive kinases. CK1 not only act as a priming kinase but also proposed as a processive kinase (**Figure 2**). Here, we focus on GSK3.

GSK3 was originally identified as a protein kinase that negatively regulates glycogen synthesis by phosphorylating and inactivating glycogen synthase [98]. In mammals, two GSK3 isoforms (GSK3α and GSK3β) are encoded by distinct genes [99]. These two GSK3 isoforms, which are expressed ubiquitously in tissues, have many overlapping functions, but they do not always compensate for each other.

Substrate recognition by GSK3 is an unusual preference for target proteins that are priorly phosphorylated at an approximately 4 residues C-terminal to GSK3 target site. The canonical consensus sequence for GSK3 is shown in **Figure 2**. Although priming phosphorylation is not stringently required for the recognition of GSK3, the efficiency of substrate phosphorylation is greatly increased by priming phosphorylation [100]. This substrate recognition mechanism means that GSK3 reduces crosstalk among different signaling pathways. In other words, GSK3 must be colocalized with the priming kinase that is involved in the specific signaling pathway. For example, in NFAT pathway, processive kinases GSK3 and CK1 are distributed throughout the entire cell. However, priming kinases DYRK1A and DYRK2 are localized to specific location, which is nuclear and cytoplasm. GSK3 thus has an unusual ability to accurately regulate the wide variety of cellular processes. We now know that the enzyme is a key regulator of various cellular processes, such as Wnt signaling pathway, hedgehog signaling pathway, NFAT pathway, mTOR pathway, EMT, cell cycle and proliferation regulation. Large number of proteins involved in a wide spectrum of cellular processes have been proposed as putative substrates of GSK3 [93]. It is noteworthy that the consensus sequence of GSK3 is included in the "phosphodegron motif" that is recognized by SCF ubiquitin-ligase complex (**Figure 4**). Therefore, most GSK3-targets receive the proteasomal degradation that relies on a phosphodegron created by dual phosphorylation and mediated by priming kinase and GSK3 [101, 102].

Unlike most protein kinases, GSK3 is constitutively active in resting cells, and its activity can be inhibited by a variety of extracellular signals that typically induce cell survival and growth, such as insulin, growth factors and nutrients. Numerous stimuli lead to activate the GSK3 inactivating kinase pathways, such as PI3K-Akt and mTOR pathway. These kinases lead to inactivation of GSK3 through the phosphorylation of N-terminal serine residue. Phosphorylated N-terminal segment creates a primed pseudosubstrate that intramolecularly binds to substrate-binding pocket (**Figure 5**). Inactivation mechanism of GSK3 by the Wnt signaling pathway is mentioned below.

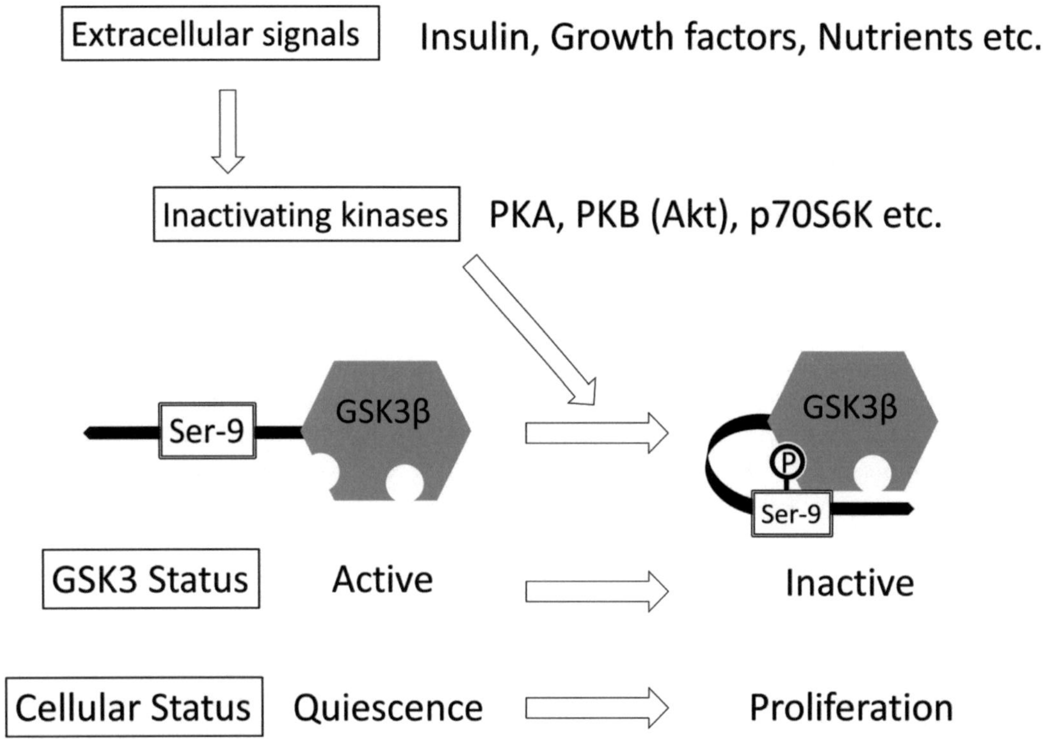

Figure 5. Regulation of GSK3 activity and cellular status. GSK3 is constitutively active in quiescent cells. Stimulation of cells with insulin, growth factors or nutrients causes inactivation of GSK3 through several kinases belonging to each signaling cascade. These GSK3 inactivating kinases phosphorylate the N-terminal serine residue of GSK3 and create a primed pseudosubstrate that binds to catalytic site and inhibits the kinase activity.

4. Priming phosphorylation regulates cellular processes

4.1. Wnt signaling pathway

Wnt signaling pathway plays crucial roles in proliferation and differentiation of stem and progenitor cells during embryogenesis and adult tissue homeostasis [103–105]. Aberration of this signaling is implicated in a variety of human cancers [106–109].

The β-catenin-dependent Wnt pathway is commonly referred to as the canonical pathway, which is characterized by the stabilization and the nuclear translocation of transcriptional co-activator β-catenin.

In the absence of Wnt ligand, β-catenin is sequestered in the cytoplasm and constantly degraded by the action of a "destruction complex," which is composed of adenomatous polyposis coli (Apc), Axin, CK1α and GSK3 [110–114]. The degradation of β-catenin is based on two steps regulated by priming phosphorylation. (1) CK1α mediates the priming phosphorylation of Apc that leads to sequential phosphorylation by GSK3, and this phosphorylation enhances the binding affinity to β-catenin [115–118]. (2) CK1α leads to phosphorylation of β-catenin on Ser-45, which creates a priming site for GSK3 [119]. Then, GSK3 phosphorylates Thr-41, Ser-37, and Ser-33 in a sequential manner [120, 121]. Priming-dependent phosphorylation by GSK3 generates the consensus motif of β-transducin repeat-containing protein (β-TRCP) recognition site at the N-terminal domain of β-catenin. After being released from the destruction complex, phosphorylated β-catenin is recognized by $SCF^{\beta\text{-TRCP}}$ E3 ubiquitin ligase and degraded by the ubiquitin-proteasome pathway [122–125] (**Figure 6**).

From a conditional knockout mouse model, the deficient of CK1α in gut epithelium shows a lot of the features of human colorectal tumors in addition to β-catenin stabilization and strong Wnt signal activation [126]. Additionally, a genome-wide, reporter-based, screening in human haploid cells reveal that CK1α is a critical negative regulator of canonical Wnt signaling pathway [127].

In the presence of Wnt ligand, β-catenin is stabilized by escaping from phosphorylation-mediated degradation and is translocated into the nucleus. After that, it binds to the T cell factor/lymphoid enhancer-binding factor (TCF/LEF) transcription factor and activates Wnt target gene expression. At this time, the function of destruction complex is suppressed by the priming phosphorylation-dependent manner (**Figure 6**).

Signaling of Wnt family requires the cell-surface receptors, frizzled (Fzd) that is related to the GPCR superfamily and low-density lipoprotein receptor-related protein 5/6 (LRP5/6) that is a single-span transmembrane receptor [128–131]. Wnt-Fzd-LRP5/6 triple complex recruits a Fzd-associated scaffold protein, Dishevelled (Dvl) and triggers the membrane-associated clustering into ribosome-sized LRP signalosomes [132]. In turn, Dvl promotes phosphorylation of the cytoplasmic tail of LRP5/6 mediated by membrane-bound CK1γ and the phosphorylated LRP5/6 is followed by the recruitment of Axin away from the degradation complex [133–136]. The phosphorylation sites of LRP5/6 contain five PPPSPxS motifs. Membrane-associated GSK3 phosphorylates the first Ser (or Thr) within these motifs and serves a priming site for the CK1-mediated phosphorylation [137, 138]. In this case, membrane-associated GSK3 acts as a priming kinase and CK1 acts as a processive kinase. The phosphorylated PPPSPxS repeats provide an optical-binding site for Axin and recruit cytoplasmic Axin-GSK3 complex to LRP signalosome [135, 137, 139]. Importantly, phosphorylated LRP5/6 cytoplasmic tail that creates a primed pseudosubstrate can directly inhibit GSK3 activity [138, 140–142], suggesting that LRP signalosome formation may be an aggressive mechanism for sequestration of GSK3 activity from cytosol. Endocytosed signalosomes that colocalize with the late endosomal markers Rab7 and Vps4 mature into multivesicular bodies [143]. As a result of the LPR signalosomes formation, β-catenin protects from phosphorylation and escapes from ubiquitylation and proteasome degradation, which enables it to accumulate in the cytosol and nucleus. As it turns out, stabilized β-catenin binds TCF/LEF to initiate the cellular transcriptional program that is usually directs to proliferation, survival and inhibition of differentiation (**Figure 7**).

4.2. Priming phosphorylation regulates protein stability

The ubiquitin-proteasome system controls degradation of the majority of regulatory proteins, including transcription factors and protein kinases, that play key roles in tumorigenesis.

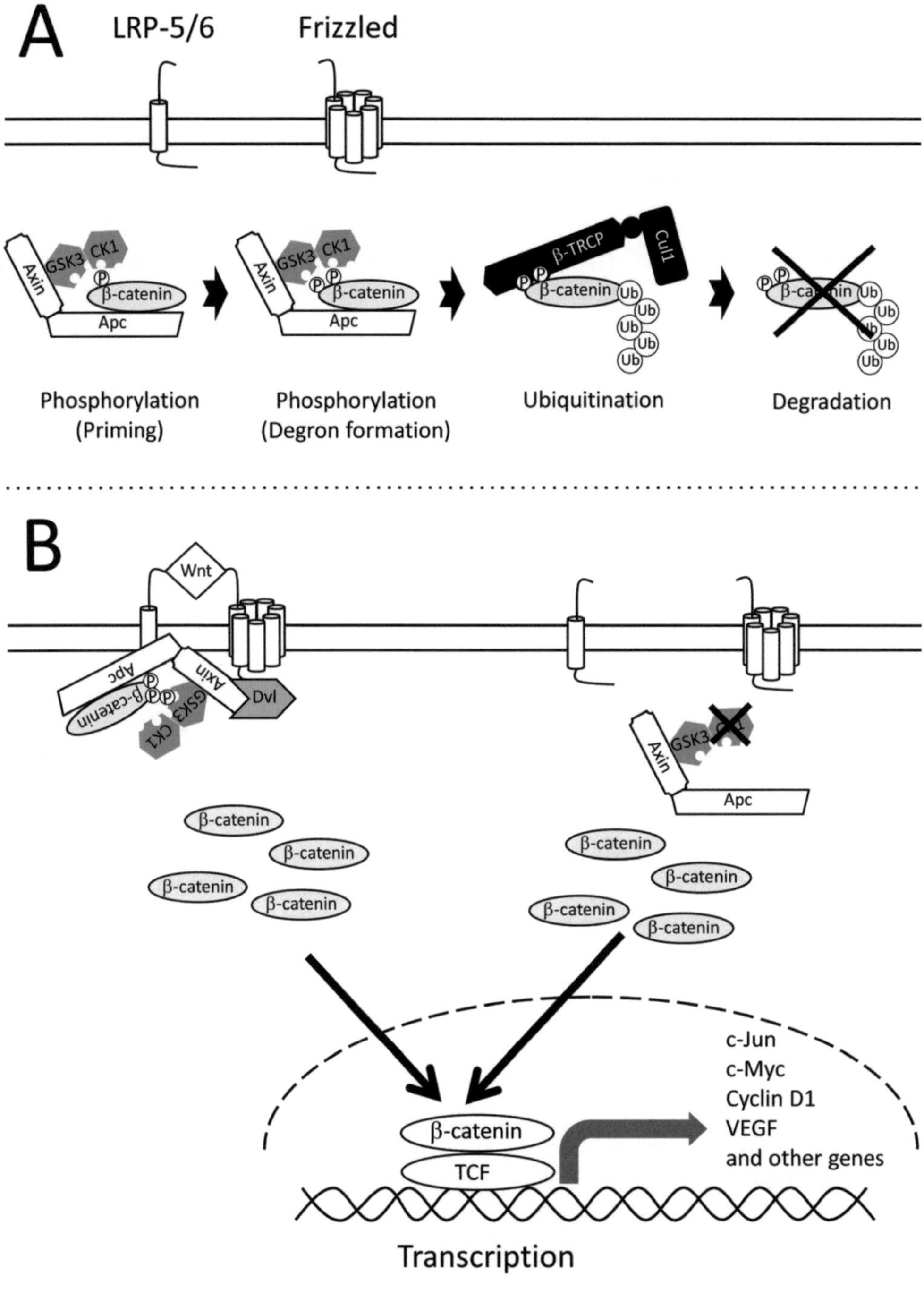

Figure 6. Canonical Wnt signaling pathway. (A) Schematic representation of constitutive degradation of β-catenin mediated by the destruction complex and $SCF^{\beta\text{-TRCP}}$ complex in resting cell. (B) Upon Wnt signaling, β-catenin is stabilized by sequestration and inactivation of destruction complex at cell membrane. It is known that the deficient of CK1α in gut epithelium shows a lot of the features of human colorectal tumors in addition to β-catenin stabilization and strong Wnt signal activation.

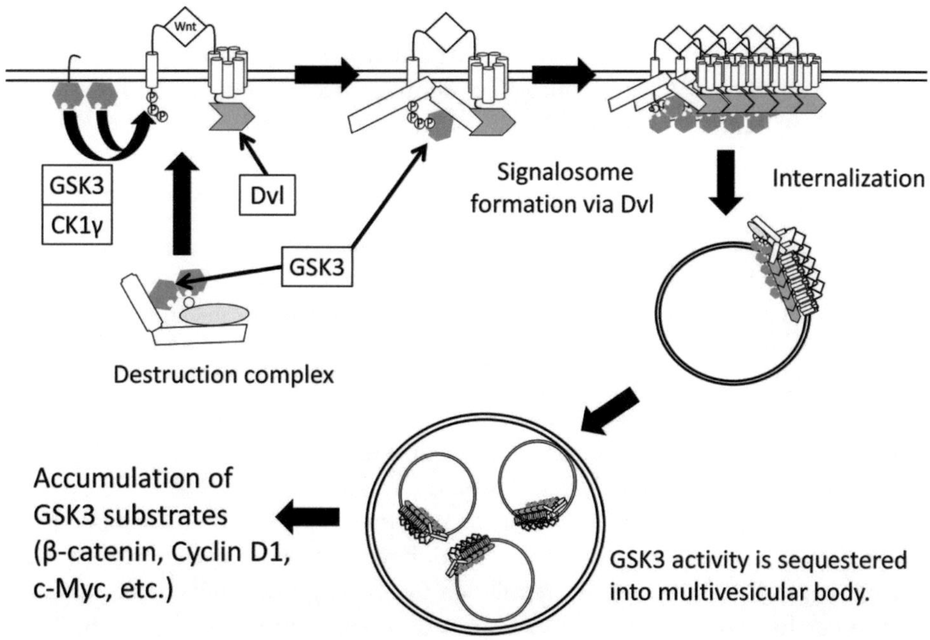

Figure 7. Wnt signaling stabilizes the GSK3 target proteins. Wnt signaling promotes the signalosome formation via priming phosphorylation of C-terminal tail of LRP5/6 and oligomerization of Dvl. The signalosome sequesters a fraction of cytosolic GSK3 into multivesicular bodies. Consequently, accumulation of GSK3 substrates including oncogenic proteins is caused by a loss of phosphodegron formation.

E3 ubiquitin ligases determine the substrate specificity for given substrates. SCF E3 ubiquitin ligase is important for the recognition of specifically phosphorylated substrates. F-box protein, which is a subunit of SCF, mediates the recognition and binding of the phosphorylated substrate. In most cases, phosphorylated substrates have a short motif that is a recognition signal for F-box proteins, namely phosphodegron. Since the consensus sequence of GSK3 is corresponding to special phosphodegron motifs that are recognized by two subfamily of F-box protein FBW7 and β-TRCP, a lot of GSK3 targets phosphorylated by priming kinases receive the proteasomal degradation [93, 102]. It is known that FBW7 and β-TRCP are involved in cell cycle regulation and tumorigenesis by targeting proteins in these processes. Thus, priming phosphorylation to create the phosphodegron processed by GSK3 is presumed to be significant consequence for cancer development.

FBW7 is generally considered as a tumor suppressor because of its loss of function phenotype found in multiple type of human cancer. FBW7 recognizes a lot of oncogenic substrates, including cyclin E, c-Myc, c-Jun, Mcl-I, mTOR and Notch-1 [144]. Among these substrates, c-Myc, c-Jun and potentially mTOR are phosphorylated by DYRK2, which creates a priming site for GSK3 [89, 145]. Therefore, the loss of priming phosphorylation may denote the direction of cancer progression in the GSK3-FBW7 axis. On the other hand, β-TRCP contributes to the degradation of β-catenin and snail, which is implicated in human cancer progression. Of note, both are phosphorylated by priming kinases [146]. In contrast to FBW7, β-TRCP is regarded as an oncogene on account of the fact that higher expression of β-TRCP is validated in various type of human cancer. Furthermore, overexpression of β-TRCP exerts its tumorigenic activity in mouse model [147] and mutations in β-TRCP are uncommon in human cancers [144]. However, due to the fact that β-TRCP substrates include both oncogene products such as β-catenin and tumor suppressors such as IκB, an inhibitor of NF-κB, it is difficult to characterize the function of β-TRCP as an oncogene or a tumor suppressor. In this context, the contribution of β-TRCP to tumor progression may become altered in the tissue specific- or cellular context-dependent manner.

Bioinformatic analysis reveals that about 20% of the human proteome contains three or more putative GSK3 phosphorylation sites. As mentioned above, LRP6 signalosome induced by the canonical Wnt signal sequesters GSK3 into multivesicular bodies and the sequestration results in the cytosolic GSK3 activity level decreased to below 40%. Accordingly, the half-life of numerous cellular proteins including GSK3 substrates is extended [143]. In proliferating cells, Wnt signaling peaks in the G2/M phase of cell cycle, and in this phase, G1 activators such as cyclin D1 and c-Myc are accumulated to progress the cell cycle [148, 149]. Moreover, depletion of GSK3 activity with a chemical inhibitor treatments or siRNA knockdown experiments stabilizes cellular proteins as similar to Wnt treatment [143]. This means that GSK3-dependent protein catabolism is more universal, beyond the cell cycle and Wnt signaling. Therefore, it is predicted that the dysregulation of priming phosphorylation influences the cellular protein homeostasis through the processive phosphorylation by GSK3. Priming kinase-GSK3-SCF axis emerges as a principal regulator of cancer development.

5. Conclusion

Hierarchical multisite phosphorylation by a priming kinase and a processive kinase is the fail-safe mechanism that accurately regulates the physiological processes, including cell cycle progression, survival, migration, metabolism, differentiation and stem cell renewal. Loss of priming phosphorylation caused by impairment of priming kinases, such as CK1 family and DYRK family, gives rise to pathological disorders as a result of the abnormal localization and/or half-life of cellular proteins. These priming kinases create the recognition site for further phosphorylation by the processive kinase, GSK3. The consensus sequence of GSK3 is corresponding to phosphodegron motif that is recognized by SCF ubiquitin-ligase complex. Therefore, a lot of GSK3 targets including oncogenes or tumor suppressors receive the proteasomal degradation that depends upon a phosphodegron. GSK3-dependent protein dissimilation is more universal, beyond the cell cycle and Wnt signaling. Consequently, priming kinase-GSK3-SCF axis manifests as a key regulator for cancer development.

Author details

Katsuhiko Aoki and Kiyotsugu Yoshida*

*Address all correspondence to: kyoshida@jikei.ac.jp

Department of Biochemistry, Jikei University School of Medicine, Tokyo, Japan

References

[1] Manning G, Plowman GD, Hunter T, Sudarsanam S. Evolution of protein kinase signaling from yeast to man. Trends in Biochemical Sciences. 2002;**27**(10):514-520

[2] Knippschild U, Gocht A, Wolff S, Huber N, Löhler J, Stöter M. The casein kinase 1 family: Participation in multiple cellular processes in eukaryotes. Cellular Signalling. 2005;**17**(6): 675-689

[3] Knippschild U, Krüger M, Richter J, Xu P, García-Reyes B, Peifer C, et al. The CK1 family: Contribution to cellular stress response and its role in carcinogenesis. Frontiers in Oncology. 2014;**19**(4):96. DOI: 10.3389/fonc.2014.00096

[4] Kumar R, Tao M. Multiple forms of casein kinase from rabbit erythrocytes. Biochimica et Biophysica Acta. 1975;**410**(1):87-98

[5] Flotow H, Roach PJ. Synergistic phosphorylation of rabbit muscle glycogen synthase by cyclic AMP-dependent protein kinase and casein kinase I. Implications for hormonal regulation of glycogen synthase. The Journal of Biological Chemistry. 1989;**264**(16):9126-9128

[6] Flotow H, Graves PR, Wang AQ, Fiol CJ, Roeske RW, Roach PJ. Phosphate groups as substrate determinants for casein kinase I action. The Journal of Biological Chemistry. 1990;**265**(24):14264-14269

[7] Agostinis P, Pinna LA, Meggio F, Marin O, Goris J, Vandenheede JR, Merlevede W. A synthetic peptide substrate specific for casein kinase I. FEBS Letters. 1989;**259**(1):75-78

[8] Flotow H, Roach PJ. Role of acidic residues as substrate determinants for casein kinase I. The Journal of Biological Chemistry. 1991;**266**(6):3724-3727

[9] Pulgar V, Marin O, Meggio F, Allende CC, Allende JE, Pinna LA. Optimal sequences for non-phosphate-directed phosphorylation by protein kinase CK1 (casein kinase-1)–a re-evaluation. European Journal of Biochemistry. 1999;**260**(2):520-526.

[10] Tuazon PT, Traugh JA. Casein kinase I and II--multipotential serine protein kinases: Structure, function, and regulation. Advances in Second Messenger and Phosphoprotein Research 1991;**23**:123-164

[11] Löhler J, Hirner H, Schmidt B, Kramer K, Fischer D, Thal DR, et al. Immunohistochemical characterisation of cell-type specific expression of CK1delta in various tissues of young adult BALB/c mice. PLoS One. 2009;**4**(1):e4174. DOI: 10.1371/journal.pone.0004174

[12] Utz AC, Hirner H, Blatz A, Hillenbrand A, Schmidt B, Deppert W, et al. Analysis of cell type-specific expression of CK1 epsilon in various tissues of young adult BALB/c Mice and in mammary tumors of SV40 T-Ag-transgenic mice. Journal of Histochemistry and Cytochemistry. 2010;**58**(1):1-15. DOI: 10.1369/jhc.2009.954628

[13] Price MA. CKI, there's more than one: Casein kinase I family members in Wnt and Hedgehog signaling. Genes and Development. 2006;**20**(4):399-410

[14] Cheong JK, Virshup DM. Casein kinase 1: Complexity in the family. International Journal of Biochemistry & Cell Biology. 2011;**43**(4):465-469. DOI: 10.1016/j.biocel.2010.12.004

[15] Ruzzene M, Tosoni K, Zanin S, Cesaro L, Pinna LA. Protein kinase CK2 accumulation in "oncophilic" cells: Causes and effects. Molecular and Cellular Biochemistry. 2011;**356**(1-2):5-10. DOI: 10.1007/s11010-011-0959-2

[16] Cruciat CM. Casein kinase 1 and Wnt/β-catenin signaling. Current Opinion in Cell Biology. 2014;**31**:46-55. DOI: 10.1016/j.ceb.2014.08.003

[17] Modak C, Chai J. Potential of casein kinase I in digestive cancer screening. World Journal of Gastrointestinal Oncology. 2009;**1**(1):26-33. DOI: 10.4251/wjgo.v1.i1.26

[18] Schittek B, Sinnberg T. Biological functions of casein kinase 1 isoforms and putative roles in tumorigenesis. Molecular Cancer. 2014;**13**:231. DOI: 10.1186/1476-4598-13-231

[19] Umar S, Wang Y, Morris AP, Sellin JH. Dual alterations in casein kinase I-epsilon and GSK-3beta modulate beta-catenin stability in hyperproliferating colonic epithelia. American Journal of Physiology - Gastrointestinal and Liver Physiology. 2007;**292**(2):G599-G607

[20] Brockschmidt C, Hirner H, Huber N, Eismann T, Hillenbrand A, Giamas G, et al. Anti-apoptotic and growth-stimulatory functions of CK1 delta and epsilon in ductal adenocarcinoma of the pancreas are inhibited by IC261 in vitro and in vivo. Gut. 2008;**57**(6):799-806. DOI: 10.1136/gut.2007.123695

[21] Sinnberg T, Menzel M, Kaesler S, Biedermann T, Sauer B, Nahnsen S, et al. Suppression of casein kinase 1alpha in melanoma cells induces a switch in beta-catenin signaling to promote metastasis. Cancer Research. 2010;**70**(17):6999-7009. DOI: 10.1158/0008-5472.CAN-10-0645

[22] Sinnberg T, Wang J, Sauer B, Schittek B. Casein kinase 1α has a non-redundant and dominant role within the CK1 family in melanoma progression. BMC Cancer 2016;**16**:594. DOI: 10.1186/s12885-016-2643-0

[23] Becker W, Joost HG. Structural and functional characteristics of Dyrk, a novel subfamily of protein kinases with dual specificity. Progress in Nucleic Acid Research and Molecular Biology 1999;**62**:1-17

[24] Kannan N, Neuwald AF. Evolutionary constraints associated with functional specificity of the CMGC protein kinases MAPK, CDK, GSK, SRPK, DYRK, and CK2alpha. Protein Science. 2004;**13**(8):2059-2077

[25] Aranda S, Laguna A, de la Luna S. DYRK family of protein kinases: Evolutionary relationships, biochemical properties, and functional roles. FASEB Journal. 2011;**25**(2):449-462. DOI: 10.1096/fj.10-165837

[26] Kentrup H, Becker W, Heukelbach J, Wilmes A, Schürmann A, Huppertz C, et al. Dyrk, a dual specificity protein kinase with unique structural features whose activity is dependent on tyrosine residues between subdomains VII and VIII. The Journal of Biological Chemistry. 1996;**271**(7):3488-3495

[27] Himpel S, Panzer P, Eirmbter K, Czajkowska H, Sayed M, Packman LC, et al. Identification of the autophosphorylation sites and characterization of their effects in the protein kinase DYRK1A. Biochemical Journal. 2001;**359**(Pt 3):497-505

[28] Lochhead PA, Sibbet G, Morrice N, Cleghon V. Activation-loop autophosphorylation is mediated by a novel transitional intermediate form of DYRKs. Cell. 2005;**121**(6):925-936

[29] Kinstrie R, Luebbering N, Miranda-Saavedra D, Sibbet G, Han J, Lochhead PA, Cleghon V. Characterization of a domain that transiently converts class 2 DYRKs into intramolecular tyrosine kinases. Science Signaling. 2010;**3**(111):ra16. DOI: 10.1126/scisignal.2000579

[30] Han J, Miranda-Saavedra D, Luebbering N, Singh A, Sibbet G, Ferguson MA, Cleghon V. Deep evolutionary conservation of an intramolecular protein kinase activation mechanism. PLoS One. 2012;**7**(1):e29702. DOI: 10.1371/journal.pone.0029702

[31] Himpel S, Tegge W, Frank R, Leder S, Joost HG, Becker W. Specificity determinants of substrate recognition by the protein kinase DYRK1A. The Journal of Biological Chemistry. 2000;**275**(4):2431-2438

[32] Campbell LE, Proud CG. Differing substrate specificities of members of the DYRK family of arginine-directed protein kinases. FEBS Letters. 2002;**510**(1-2):31-36

[33] Papadopoulos C, Arato K, Lilienthal E, Zerweck J, Schutkowski M, Chatain N, et al. Splice variants of the dual specificity tyrosine phosphorylation-regulated kinase 4 (DYRK4) differ in their subcellular localization and catalytic activity. The Journal of Biological Chemistry. 2011;**286**(7):5494-5505. DOI: 10.1074/jbc.M110.157909

[34] Soundararajan M, Roos AK, Savitsky P, Filippakopoulos P, Kettenbach AN, Olsen JV, et al. Structures of Down syndrome kinases, DYRKs, reveal mechanisms of kinase activation and substrate recognition. Structure. 2013;**21**(6):986-996. DOI: 10.1016/j.str.2013.03.012

[35] Malinge S, Bliss-Moreau M, Kirsammer G, Diebold L, Chlon T, Gurbuxani S, Crispino JD. Increased dosage of the chromosome 21 ortholog Dyrk1a promotes megakaryoblastic leukemia in a murine model of Down syndrome. Journal of Clinical Investigation. 2012;**122**(3):948-962. DOI: 10.1172/JCI60455

[36] Pozo N, Zahonero C, Fernández P, Liñares JM, Ayuso A, Hagiwara M, et al. Inhibition of DYRK1A destabilizes EGFR and reduces EGFR-dependent glioblastoma growth. Journal of Clinical Investigation. 2013;**123**(6):2475-2487. DOI: 10.1172/JCI63623

[37] Guo X, Wang X, Wang Z, Banerjee S, Yang J, Huang L, Dixon JE. Site-specific proteasome phosphorylation controls cell proliferation and tumorigenesis. Nature Cell Biology. 2016;**18**(2):202-212. DOI: 10.1038/ncb3289

[38] Baek KH, Zaslavsky A, Lynch RC, Britt C, Okada Y, Siarey RJ, et al. Down's syndrome suppression of tumour growth and the role of the calcineurin inhibitor DSCR1. Nature. 2009;**459**(7250):1126-1130. DOI: 10.1038/nature08062

[39] Litovchick L, Florens LA, Swanson SK, Washburn MP, DeCaprio JA. DYRK1A protein kinase promotes quiescence and senescence through DREAM complex assembly. Genes and Development. 2011;**25**(8):801-813. DOI: 10.1101/gad.2034211

[40] Tschöp K, Conery AR, Litovchick L, Decaprio JA, Settleman J, Harlow E, Dyson N. A kinase shRNA screen links LATS2 and the pRB tumor suppressor. Genes and Development. 2011;**25**(8):814-830. DOI: 10.1101/gad.2000211

[41] Liu Q, Liu N, Zang S, Liu H, Wang P, Ji C, Sun X. Tumor suppressor DYRK1A effects on proliferation and chemoresistance of AML cells by downregulating c-Myc. PLoS One. 2014;**9**(6):e98853. DOI: 10.1371/journal.pone.0098853

[42] Yamashita S, Chujo M, Moroga T, Anami K, Tokuishi K, Miyawaki M, et al. DYRK2 expression may be a predictive marker for chemotherapy in non-small cell lung cancer. Anticancer Research. 2009;**29**(7):2753-2757

[43] Yamashita S, Chujo M, Tokuishi K, Anami K, Miyawaki M, Yamamoto S, Kawahara K. Expression of dual-specificity tyrosine-(Y)-phosphorylation-regulated kinase 2 (DYRK2) can be a favorable prognostic marker in pulmonary adenocarcinoma. Journal of Thoracic and Cardiovascular Surgery. 2009;**138**(6):1303-1308. DOI: 10.1016/j.jtcvs.2009.08.003

[44] Park JH, Park J, Choi JK, Lyu J, Bae MG, Lee YG, et al. Identification of DNA methylation changes associated with human gastric cancer. BMC Medical Genomics 2011;**4**:82. DOI: 10.1186/1755-8794-4-82

[45] Enomoto Y, Yamashita S, Yoshinaga Y, Fukami Y, Miyahara S, Nabeshima K, Iwasaki A. Downregulation of DYRK2 can be a predictor of recurrence in early stage breast cancer. Tumour Biology. 2014;**35**(11):11021-11025. DOI: 10.1007/s13277-014-2413-z

[46] Nomura S, Suzuki Y, Takahashi R, Terasaki M, Kimata R, Terasaki Y, et al. Dual-specificity tyrosine phosphorylation-regulated kinase 2 (DYRK2) as a novel marker in T1 high-grade and T2 bladder cancer patients receiving neoadjuvant chemotherapy. BMC Urology. 2015;**15**:53. DOI: 10.1186/s12894-015-0040-7

[47] Wang Y, Wu Y, Miao X, Zhu X, Miao X, He Y, et al. Silencing of DYRK2 increases cell proliferation but reverses CAM-DR in Non-Hodgkin's Lymphoma. International Journal of Biological Macromolecules 2015;**81**:809-817. DOI: 10.1016/j.ijbiomac.2015.08.067

[48] Zhang X, Xu P, Ni W, Fan H, Xu J, Chen Y, et al. Downregulated DYRK2 expression is associated with poor prognosis and oxaliplatin resistance in hepatocellular carcinoma. Pathology, Research and Practice. 2016;**212**(3):162-170. DOI: 10.1016/j.prp.2016.01.002

[49] Yan H, Hu K, Wu W, Li Y, Tian H, Chu Z, et al. Low expression of DYRK2 (Dual specificity tyrosine phosphorylation regulated kinase 2) correlates with poor prognosis in colorectal cancer. PLoS One. 2016;**11**(8):e0159954. DOI: 10.1371/journal.pone.0159954

[50] Friedman E. Mirk/Dyrk1B in cancer. Journal of Cellular Biochemistry. 2007;**102**(2):274-279

[51] Jin K, Park S, Ewton DZ, Friedman E. The survival kinase Mirk/Dyrk1B is a downstream effector of oncogenic K-ras in pancreatic cancer. Cancer Research. 2007;**67**(15):7247-7255

[52] Deng X, Ewton DZ, Friedman E. Mirk/Dyrk1B maintains the viability of quiescent pancreatic cancer cells by reducing levels of reactive oxygen species. Cancer Research. 2009;**69**(8):3317-3324. DOI: 10.1158/0008-5472.CAN-08-2903

[53] Jin K, Ewton DZ, Park S, Hu J, Friedman E. Mirk regulates the exit of colon cancer cells from quiescence. The Journal of Biological Chemistry. 2009;**284**(34):22916-22925. DOI: 10.1074/jbc.M109.035519

[54] Gao J, Zheng Z, Rawal B, Schell MJ, Bepler G, Haura EB. Mirk/Dyrk1B, a novel therapeutic target, mediates cell survival in non-small cell lung cancer cells. Cancer Biology and Therapy. 2009;**8**(17):1671-1679

[55] Yang C, Ji D, Weinstein EJ, Choy E, Hornicek FJ, Wood KB, et al. The kinase Mirk is a potential therapeutic target in osteosarcoma. Carcinogenesis. 2010;**31**(4):552-558. DOI: 10.1093/carcin/bgp330

[56] Hu J, Nakhla H, Friedman E. Transient arrest in a quiescent state allows ovarian cancer cells to survive suboptimal growth conditions and is mediated by both Mirk/dyrk1b and p130/RB2. International Journal of Cancer. 2011;**129**(2):307-318. DOI: 10.1002/ijc.25692

[57] Ewton DZ, Hu J, Vilenchik M, Deng X, Luk KC, Polonskaia A, et al. Inactivation of mirk/dyrk1b kinase targets quiescent pancreatic cancer cells. Molecular Cancer Therapeutics. 2011;**10**(11):2104-2114. DOI: 10.1158/1535-7163.MCT-11-0498

[58] Gao J, Yang X, Yin P, Hu W, Liao H, Miao Z, et al. The involvement of FoxO in cell survival and chemosensitivity mediated by Mirk/Dyrk1B in ovarian cancer. International Journal of Oncology. 2012;**40**(4):1203-1209. DOI: 10.3892/ijo.2011.1293

[59] Hu J, Deng H, Friedman EA. Ovarian cancer cells, not normal cells, are damaged by Mirk/Dyrk1B kinase inhibition. International Journal of Cancer. 2013;**132**(10):2258-2269. DOI: 10.1002/ijc.27917

[60] Tsioras K, Papastefanaki F, Politis PK, Matsas R, Gaitanou M. Functional Interactions between BM88/Cend1, Ran-binding protein M and Dyrk1B kinase affect cyclin D1 levels and cell cycle progression/exit in mouse neuroblastoma cells. PLoS One. 2013;**8**(11):e82172. DOI: 10.1371/journal.pone.0082172

[61] Gruber W, Hutzinger M, Elmer DP, Parigger T, Sternberg C, Cegielkowski L, et al. DYRK1B as therapeutic target in Hedgehog/GLI-dependent cancer cells with Smoothened inhibitor resistance. Oncotarget. 2016;**7**(6):7134-7148. DOI: 10.18632/oncotarget.6910

[62] Fernández-Martínez P, Zahonero C, Sánchez-Gómez P. DYRK1A: The double-edged kinase as a protagonist in cell growth and tumorigenesis. Molecular & Cellular Oncology. 2015;**2**(1):e970048. DOI: 10.4161/23723548.2014.970048

[63] Abbassi R, Johns TG, Kassiou M, Munoz L. DYRK1A in neurodegeneration and cancer: Molecular basis and clinical implications. Pharmacology and Therapeutics. 2015;**151**:87-98. DOI: 10.1016/j.pharmthera.2015.03.004

[64] Shindoh N, Kudoh J, Maeda H, Yamaki A, Minoshima S, Shimizu Y, Shimizu N. Cloning of a human homolog of the Drosophila minibrain/rat Dyrk gene from "the Down syndrome critical region" of chromosome 21. Biochemical and Biophysical Research Communications. 1996;**225**(1):92-99

[65] Guimerá J, Casas C, Pucharcòs C, Solans A, Domènech A, Planas AM, et al. A human homologue of Drosophila minibrain (MNB) is expressed in the neuronal regions affected in Down syndrome and maps to the critical region. Human Molecular Genetics. 1996;**5**(9):1305-1310

[66] Song WJ, Sternberg LR, Kasten-Sportès C, Keuren ML, Chung SH, Slack AC, et al. Isolation of human and murine homologues of the Drosophila minibrain gene: Human homologue maps to 21q22.2 in the Down syndrome "critical region". Genomics. 1996;**38**(3):331-339

[67] Matsumoto N, Ohashi H, Tsukahara M, Kim KC, Soeda E, Niikawa N. Possible narrowed assignment of the loci of monosomy 21-associated microcephaly and intrauterine growth retardation to a 1.2-Mb segment at 21q22.2. The American Journal of Human Genetic. 1997;**60**(4):997-999

[68] Smith DJ, Stevens ME, Sudanagunta SP, Bronson RT, Makhinson M, Watabe AM, et al. Functional screening of 2 Mb of human chromosome 21q22.2 in transgenic mice implicates minibrain in learning defects associated with Down syndrome. Nature Genetics. 1997;**16**(1):28-36

[69] Guimera J, Casas C, Estivill X, Pritchard M. Human minibrain homologue (MNBH/DYRK1): Characterization, alternative splicing, differential tissue expression, and overexpression in Down syndrome. Genomics. 1999;**57**(3):407-418

[70] Altafaj X, Dierssen M, Baamonde C, Martí E, Visa J, Guimerà J, et al. Neurodevelopmental delay, motor abnormalities and cognitive deficits in transgenic mice overexpressing Dyrk1A (minibrain), a murine model of Down's syndrome. Human Molecular Genetics. 2001;**10**(18):1915-1923

[71] Fotaki V, Dierssen M, Alcántara S, Martínez S, Martí E, Casas C, et al. Dyrk1A haploinsufficiency affects viability and causes developmental delay and abnormal brain morphology in mice. Molecular Cell. Biology. 2002;**22**(18):6636-6647

[72] Laguna A, Aranda S, Barallobre MJ, Barhoum R, Fernández E, Fotaki V, et al. The protein kinase DYRK1A regulates caspase-9-mediated apoptosis during retina development. Developmental Cell. 2008;**15**(6):841-853. DOI: 10.1016/j.devcel.2008.10.014

[73] Yabut O, Domogauer J, D'Arcangelo G. Dyrk1A overexpression inhibits proliferation and induces premature neuronal differentiation of neural progenitor cells. Journal of Neuroscience. 2010;**30**(11):4004-4014. DOI: 10.1523/JNEUROSCI.4711-09.2010

[74] Guedj F, Pereira PL, Najas S, Barallobre MJ, Chabert C, Souchet B, et al. DYRK1A: A master regulatory protein controlling brain growth. Neurobiology of Disease. 2012;**46**(1):190-203. DOI: 10.1016/j.nbd.2012.01.007

[75] Duchon A, Herault Y. DYRK1A, a dosage-sensitive gene involved in neurodevelopmental disorders, is a target for drug development in Down syndrome. Frontiers in Behavioral Neuroscience 2016;**10**:104. DOI: 10.3389/fnbeh.2016.00104

[76] Park J, Oh Y, Yoo L, Jung MS, Song WJ, Lee SH, et al. Dyrk1A phosphorylates p53 and inhibits proliferation of embryonic neuronal cells. The Journal of Biological Chemistry. 2010;**285**(41):31895-31906. DOI: 10.1074/jbc.M110.147520

[77] Hasle H, Clemmensen IH, Mikkelsen M. Risks of leukaemia and solid tumours in individuals with Down's syndrome. Lancet. 2000;**355**(9199):165-169

[78] Hasle H. Pattern of malignant disorders in individuals with Down's syndrome. The Lancet Oncology. 20011;**2**(7):429-436

[79] Nižetić D, Groet J. Tumorigenesis in Down's syndrome: Big lessons from a small chromosome. Nature Reviews Cancer. 2012;**12**(10):721-732. DOI: 10.1038/nrc3355

[80] Mancini M, Toker A. NFAT proteins: Emerging roles in cancer progression. Nature Reviews Cancer. 2009;**9**(11):810-820. DOI: 10.1038/nrc2735

[81] Müller MR, Rao A. NFAT, immunity and cancer: A transcription factor comes of age. Nature Reviews Immunology. 2010;**10**(9):645-656. DOI: 10.1038/nri2818

[82] Shou J, Jing J, Xie J, You L, Jing Z, Yao J, et al. Nuclear factor of activated T cells in cancer development and treatment. Cancer Letters. 2015;**361**(2):174-184. DOI: 10.1016/j.canlet.2015.03.005

[83] Gwack Y, Sharma S, Nardone J, Tanasa B, Iuga A, Srikanth S, et al. A genome-wide Drosophila RNAi screen identifies DYRK-family kinases as regulators of NFAT. Nature. 2006;**441**(7093):646-650

[84] Taira N, Nihira K, Yamaguchi T, Miki Y, Yoshida K. DYRK2 is targeted to the nucleus and controls p53 via Ser46 phosphorylation in the apoptotic response to DNA damage. Molecular Cell. 2007;**25**(5):725-738

[85] Taira N, Yamamoto H, Yamaguchi T, Miki Y, Yoshida K. ATM augments nuclear stabilization of DYRK2 by inhibiting MDM2 in the apoptotic response to DNA damage. The Journal of Biological Chemistry. 2010;**285**(7):4909-4919. DOI: 10.1074/jbc.M109.042341

[86] Yamamoto T, Nihira NT, Yogosawa S, Aoki K, Takeda H, Sawasaki T, Yoshida K. Interaction between RNF8 and DYRK2 is required for the recruitment of DNA repair molecules to DNA double-strand breaks. FEBS Letters. 2017;**591**(6):842-853. DOI: 10.1002/1873-3468.12596

[87] Stephens P, Edkins S, Davies H, Greenman C, Cox C, Hunter C, et al. A screen of the complete protein kinase gene family identifies diverse patterns of somatic mutations in human breast cancer. Nature Genetics. 2005;**37**(6):590-592

[88] Greenman C, Stephens P, Smith R, Dalgliesh GL, Hunter C, Bignell G, et al. Patterns of somatic mutation in human cancer genomes. Nature. 2007;**446**(7132):153-158

[89] Taira N, Mimoto R, Kurata M, Yamaguchi T, Kitagawa M, Miki Y, Yoshida K. DYRK2 priming phosphorylation of c-Jun and c-Myc modulates cell cycle progression in human cancer cells. Journal of Clinical Investigation. 2012;**122**(3):859-872. DOI: 10.1172/JCI60818

[90] Mimoto R, Taira N, Takahashi H, Yamaguchi T, Okabe M, Uchida K, et al. DYRK2 controls the epithelial-mesenchymal transition in breast cancer by degrading Snail. Cancer Letters. 2013;**339**(2):214-225. DOI: 10.1016/j.canlet.2013.06.005

[91] Yamaguchi N, Mimoto R, Yanaihara N, Imawari Y, Hirooka S, Okamoto A, Yoshida K. DYRK2 regulates epithelial-mesenchymal-transition and chemosensitivity through Snail degradation in ovarian serous adenocarcinoma. Tumour Biology. 2015;**36**(8):5913-5923. DOI: 10.1007/s13277-015-3264-y

[92] Mimoto R, Imawari Y, Hirooka S, Takeyama H, Yoshida K. Impairment of DYRK2 augments stem-like traits by promoting KLF4 expression in breast cancer. Oncogene. 2017;36(13):1862-1872. DOI: 10.1038/onc.2016.349

[93] Sutherland C. What are the bona fide GSK3 substrates? International Journal of Alzheimer's Disease. 2011;**2011**:505607. DOI: 10.4061/2011/505607

[94] Park JE, Soung NK, Johmura Y, Kang YH, Liao C, Lee KH, et al. Polo-box domain: A versatile mediator of polo-like kinase function. Cellular and Molecular Life Sciences. 2010;**67**(12):1957-1970. DOI: 10.1007/s00018-010-0279-9

[95] Yuan K, Huang Y, Yao X. Illumination of mitotic orchestra during cell division: A Polo view. Cellular Signalling. 2011;**23**(1):1-5. DOI:10.1016/j.cellsig.2010.07.003

[96] Meggio F, Pinna LA. One-thousand-and-one substrates of protein kinase CK2? FASEB Journal. 2003;**17**(3):349-368

[97] St-Denis N, Gabriel M, Turowec JP, Gloor GB, Li SS, Gingras AC, Litchfield DW. Systematic investigation of hierarchical phosphorylation by protein kinase CK2. Journal of Proteomics. 2015;**118**:49-62. DOI: 10.1016/j.jprot.2014.10.020

[98] Embi N, Rylatt DB, Cohen P. Glycogen synthase kinase-3 from rabbit skeletal muscle. Separation from cyclic-AMP-dependent protein kinase and phosphorylase kinase. European Journal of Biochemistry. 1980;**107**(2):519-527

[99] Woodgett JR. Molecular cloning and expression of glycogen synthase kinase-3/factor A. The EMBO Journal. 1990;**9**(8):2431-2438

[100] Thomas GM, Frame S, Goedert M, Nathke I, Polakis P, Cohen P. A GSK3-binding peptide from FRAT1 selectively inhibits the GSK3-catalysed phosphorylation of axin and beta-catenin. FEBS Letters. 1999;**458**(2):247-251

[101] Kim NG, Xu C, Gumbiner BM. Identification of targets of the Wnt pathway destruction complex in addition to beta-catenin. Proceedings of the National Academy of Sciences of the United States of America. 2009;**106**(13):5165-5170. DOI: 10.1073/pnas.0810185106

[102] Xu C, Kim NG, Gumbiner BM. Regulation of protein stability by GSK3 mediated phosphorylation. Cell Cycle. 2009;**8**(24):4032-4039

[103] Wodarz A, Nusse R. Mechanisms of Wnt signaling in development. Annual Review of Cell and Developmental Biology 1998;**14**:59-88

[104] Clevers H, Nusse R. Wnt/β-catenin signaling and disease. Cell. 2012;**149**(6):1192-1205. DOI: 10.1016/j.cell.2012.05.012

[105] Schuijers J, Clevers H. Adult mammalian stem cells: The role of Wnt, Lgr5 and R-spondins. The EMBO Journal. 2012;**31**(12):2685-2696. DOI: 10.1038/emboj.2012.149

[106] Polakis P. The many ways of Wnt in cancer. Current Opinion in Genetics and Development. 2007;**17**(1):45-51

[107] Kandoth C, McLellan MD, Vandin F, Ye K, Niu B, Lu C, et al. Mutational landscape and significance across 12 major cancer types. Nature. 2013;**502**(7471):333-339. DOI: 10.1038/nature12634

[108] Zhan T, Rindtorff N, Boutros M. Wnt signaling in cancer. Oncogene. 2017;**36**(11):1461-1473. DOI: 10.1038/onc.2016.304

[109] Duchartre Y, Kim YM, Kahn M. The Wnt signaling pathway in cancer. Critical Reviews in Oncology/Hematology 2016;**99**:141-149. DOI: 10.1016/j.critrevonc.2015.12.005

[110] Rubinfeld B, Albert I, Porfiri E, Fiol C, Munemitsu S, Polakis P. Binding of GSK3beta to the APC-beta-catenin complex and regulation of complex assembly. Science. 1996;**272**(5264):1023-1026

[111] Ikeda S, Kishida S, Yamamoto H, Murai H, Koyama S, Kikuchi A. Axin, a negative regulator of the Wnt signaling pathway, forms a complex with GSK-3beta and beta-catenin and promotes GSK-3beta-dependent phosphorylation of beta-catenin. The EMBO Journal. 1998;**17**(5):1371-1384

[112] Nakamura T, Hamada F, Ishidate T, Anai K, Kawahara K, Toyoshima K, Akiyama T. Axin, an inhibitor of the Wnt signalling pathway, interacts with beta-catenin, GSK-3beta and APC and reduces the beta-catenin level. Genes to Cells. 1998;**3**(6):395-403

[113] Li VS, Ng SS, Boersema PJ, Low TY, Karthaus WR, Gerlach JP, et al. Wnt signaling through inhibition of β-catenin degradation in an intact Axin1 complex. Cell. 2012;**149**(6):1245-1256. DOI: 10.1016/j.cell.2012.05.002

[114] Schwarz-Romond T, Asbrand C, Bakkers J, Kühl M, Schaeffer HJ, Huelsken J, et al. The ankyrin repeat protein Diversin recruits Casein kinase Iepsilon to the beta-catenin degradation complex and acts in both canonical Wnt and Wnt/JNK signaling. Genes and Development. 2002;**16**(16):2073-2084

[115] Ha NC, Tonozuka T, Stamos JL, Choi HJ, Weis WI. Mechanism of phosphorylation-dependent binding of APC to beta-catenin and its role in beta-catenin degradation. Molecular Cell. 2004;**15**(4):511-521

[116] Xing Y, Clements WK, Le Trong I, Hinds TR, Stenkamp R, Kimelman D, Xu W. Crystal structure of a beta-catenin/APC complex reveals a critical role for APC phosphorylation in APC function. Molecular Cell. 2004;**15**(4):523-533

[117] Liu J, Xing Y, Hinds TR, Zheng J, Xu W. The third 20 amino acid repeat is the tightest binding site of APC for beta-catenin. Journal of Molecular Biology. 2006;**360**(1):133-144

[118] Ferrarese A, Marin O, Bustos VH, Venerando A, Antonelli M, Allende JE, Pinna LA. Chemical dissection of the APC Repeat 3 multistep phosphorylation by the concerted action of protein kinases CK1 and GSK3. Biochemistry. 2007;**46**(42):11902-11910

[119] Amit S, Hatzubai A, Birman Y, Andersen JS, Ben-Shushan E, Mann M, et al. Axin-mediated CKI phosphorylation of beta-catenin at Ser 45: A molecular switch for the Wnt pathway. Genes and Development. 2002;**16**(9):1066-1076

[120] Liu C, Li Y, Semenov M, Han C, Baeg GH, Tan Y, et al. Control of beta-catenin phosphorylation/degradation by a dual-kinase mechanism. Cell. 2002;**108**(6):837-847

[121] Hagen T, Vidal-Puig A. Characterisation of the phosphorylation of beta-catenin at the GSK-3 priming site Ser45. Biochemical and Biophysical Research Communications. 2002;**294**(2):324-328

[122] Marikawa Y, Elinson RP. beta-TrCP is a negative regulator of Wnt/beta-catenin signaling pathway and dorsal axis formation in Xenopus embryos. Mechanisms of Development. 1998;**77**(1):75-80

[123] Hart M, Concordet JP, Lassot I, Albert I, del los Santos R, Durand H, et al. The F-box protein beta-TrCP associates with phosphorylated beta-catenin and regulates its activity in the cell. Current Biology. 1999;**9**(4):207-210

[124] Kitagawa M, Hatakeyama S, Shirane M, Matsumoto M, Ishida N, Hattori K, et al. An F-box protein, FWD1, mediates ubiquitin-dependent proteolysis of beta-catenin. The EMBO Journal. 1999;**18**(9):2401-2410

[125] Su Y, Fu C, Ishikawa S, Stella A, Kojima M, Shitoh K, et al. APC is essential for targeting phosphorylated beta-catenin to the SCFbeta-TrCP ubiquitin ligase. Molecular Cell. 2008;**32**(5):652-661. DOI: 10.1016/j.molcel.2008.10.023

[126] Elyada E, Pribluda A, Goldstein RE, Morgenstern Y, Brachya G, Cojocaru G, et al. CKIα ablation highlights a critical role for p53 in invasiveness control. Nature. 2011;**470**(7334):409-413. DOI: 10.1038/nature09673

[127] Lebensohn AM, Dubey R, Neitzel LR, Tacchelly-Benites O, Yang E, Marceau CD, Davis EM, Patel BB, Bahrami-Nejad Z, Travaglini KJ, Ahmed Y, Lee E, Carette JE, Rohatgi R. Comparative genetic screens in human cells reveal new regulatory mechanisms in WNT signaling. Elife. 2016;**5**(pii):e21459. DOI: 10.7554/eLife.21459

[128] Bhanot P, Brink M, Samos CH, Hsieh JC, Wang Y, Macke JP, et al. A new member of the frizzled family from Drosophila functions as a Wingless receptor. Nature. 1996;**382**(6588):225-230.

[129] Wehrli M, Dougan ST, Caldwell K, O'Keefe L, Schwartz S, Vaizel-Ohayon D, et al. arrow encodes an LDL-receptor-related protein essential for Wingless signalling. Nature. 2000;**407**(6803):527-530.

[130] Tamai K, Semenov M, Kato Y, Spokony R, Liu C, Katsuyama Y, et al. LDL-receptor-related proteins in Wnt signal transduction. Nature. 2000;**407**(6803):530-535

[131] He X, Semenov M, Tamai K, Zeng X. LDL receptor-related proteins 5 and 6 in Wnt/beta-catenin signaling: Arrows point the way. Development. 2004;**131**(8):1663-1677.

[132] Bilic J, Huang YL, Davidson G, Zimmermann T, Cruciat CM, Bienz M, Niehrs C. Wnt induces LRP6 signalosomes and promotes dishevelled-dependent LRP6 phosphorylation. Science. 2007;**316**(5831):1619-1622

[133] Mao J, Wang J, Liu B, Pan W, Farr GH 3rd, Flynn C, et al. Low-density lipoprotein receptor-related protein-5 binds to Axin and regulates the canonical Wnt signaling pathway. Molecular Cell. 2001;**7**(4):801-809

[134] Cliffe A, Hamada F, Bienz M. A role of Dishevelled in relocating Axin to the plasma membrane during wingless signaling. Current Biology. 2003;**13**(11):960-966

[135] Tamai K, Zeng X, Liu C, Zhang X, Harada Y, Chang Z, He X. A mechanism for Wnt coreceptor activation. Molecular Cell. 2004;**13**(1):149-156

[136] Davidson G, Wu W, Shen J, Bilic J, Fenger U, Stannek P, et al. Casein kinase 1 gamma couples Wnt receptor activation to cytoplasmic signal transduction. Nature. 2005;**438**(7069):867-872

[137] Zeng X, Tamai K, Doble B, Li S, Huang H, Habas R, et al. A dual-kinase mechanism for Wnt co-receptor phosphorylation and activation. Nature. 2005;**438**(7069):873-877

[138] Mi K, Dolan PJ, Johnson GV. The low density lipoprotein receptor-related protein 6 interacts with glycogen synthase kinase 3 and attenuates activity. The Journal of Biological Chemistry. 2006;**281**(8):4787-4794

[139] Zeng X, Huang H, Tamai K, Zhang X, Harada Y, Yokota C, et al. Initiation of Wnt signaling: Control of Wnt coreceptor Lrp6 phosphorylation/activation via frizzled, dishevelled and axin functions. Development. 2008;**135**(2):367-375

[140] Cselenyi CS, Jernigan KK, Tahinci E, Thorne CA, Lee LA, Lee E. LRP6 transduces a canonical Wnt signal independently of Axin degradation by inhibiting GSK3's phosphorylation of beta-catenin. Proceedings of the National Academy of Sciences of the United States of America. 2008;**105**(23):8032-8037. DOI: 10.1073/pnas.0803025105

[141] Piao S, Lee SH, Kim H, Yum S, Stamos JL, Xu Y, et al. Direct inhibition of GSK3beta by the phosphorylated cytoplasmic domain of LRP6 in Wnt/beta-catenin signaling. PLoS One. 2008;**3**(12):e4046. DOI: 10.1371/journal.pone.0004046

[142] Wu G, Huang H, Garcia Abreu J, He X. Inhibition of GSK3 phosphorylation of beta-catenin via phosphorylated PPPSPXS motifs of Wnt coreceptor LRP6. PLoS One. 2009;**4**(3):e4926. DOI: 10.1371/journal.pone.0004926

[143] Taelman VF, Dobrowolski R, Plouhinec JL, Fuentealba LC, Vorwald PP, Gumper I, et al. Wnt signaling requires sequestration of glycogen synthase kinase 3 inside multivesicular endosomes. Cell. 2010;**143**(7):1136-1148. DOI: 10.1016/j.cell.2010.11.034

[144] Lau AW, Fukushima H, Wei W. The Fbw7 and betaTRCP E3 ubiquitin ligases and their roles in tumorigenesis. Frontiers in Bioscience (Landmark edition). 2012;**17**:2197-2212

[145] Mimoto R, Nihira NT, Hirooka S, Takeyama H, Yoshida K. Diminished DYRK2 sensitizes hormone receptor-positive breast cancer to everolimus by the escape from degrading mTOR. Cancer Letters. 2017;**384**:27-38. DOI: 10.1016/j.canlet.2016.10.015

[146] Díaz VM, de Herreros AG. F-box proteins: Keeping the epithelial-to-mesenchymal transition (EMT) in check. Seminars in Cancer Biology 2016;**36**:71-79. DOI: 10.1016/j.semcancer.2015.10.003

[147] Zheng N, Zhou Q, Wang Z, Wei W. Recent advances in SCF ubiquitin ligase complex: Clinical implications. Biochimica et Biophysica Acta. 2016;**1866**(1):12-22. DOI: 10.1016/j.bbcan.2016.05.001

[148] Davidson G, Shen J, Huang YL, Su Y, Karaulanov E, Bartscherer K, et al. Cell cycle control of wnt receptor activation. Developmental Cell. 2009;**17**(6):788-799. DOI: 10.1016/j.devcel.2009.11.006

[149] Acebron SP, Karaulanov E, Berger BS, Huang YL, Niehrs C. Mitotic wnt signaling promotes protein stabilization and regulates cell size. Molecular Cell. 2014;**54**(4):663-674. DOI: 10.1016/j.molcel.2014.04.014

Chapter 5

Phosphorylation-Mediated Control of Stress Responses Induced by Nanosecond Pulsed Electric Fields

Ken-ichi Yano and Keiko Morotomi-Yano

Additional information is available at the end of the chapter

http://dx.doi.org/10.5772/intechopen.69782

Abstract

Exposure of living organisms to short electric pulses is widely utilized in the life sciences, for example, for DNA transfection. Recent advances in electrical engineering have enabled the production of extremely short electric pulses in the range of nanoseconds, namely, nanosecond pulsed electric fields (nsPEFs). nsPEFs are increasingly recognized as a novel means for cancer therapy, because of their ability to induce cell death. Recent studies have demonstrated that nsPEFs act as cellular stress and activate two independent signaling pathways that involve phosphorylation of translation initiation factors and lead to suppression of general protein synthesis. eIF2α phosphorylation is one of the key reactions in stress-induced translational suppression and is rapidly induced by nsPEFs. Concomitantly, PERK and GCN2, both of which are stress-responsive protein kinases, are activated in nsPEF-exposed cells. Furthermore, nsPEFs cause a reduction in 4E-BP1 phosphorylation, which is controlled by mTORC1 and constitutes an alternative mechanism for translational suppression, independent of eIF2α phosphorylation. In accordance with elevated eIF2α phosphorylation and decreased 4E-BP1 phosphorylation, general protein synthesis is acutely suppressed after nsPEF exposure. These findings demonstrate that nsPEFs induce two independent signaling pathways for translational suppression, further highlighting a unique feature of nsPEFs as a novel means for life sciences.

Keywords: stress response, electroporation, pulsed electric field, eIF2α, PERK, GCN2, 4E-BP1, translational suppression, protein synthesis

1. Introduction

Living cells respond to various environmental stimuli by activating distinct sets of intracellular reactions. Cellular responses to external stimuli generally involve signal transduction that is mediated by protein phosphorylation and eventually leads to modulation of various cellular events, such as metabolism, gene expression, proliferation, and cell death [1].

Some physical and chemical stimuli have adverse physiological effects and are known as cellular stress. In addition to adverse external stimuli, endogenous deleterious events are also regarded as cellular stress, such as accumulation of misfolded proteins in the endoplasmic reticulum (ER). Cells respond to these cellular stresses by inducing various intracellular reactions, which are collectively referred to as stress responses. Timely induction of stress responses is critical for maintenance of cellular physiology, and its dysregulation is frequently observed in various diseases, such as cancer [2], neurodegenerative disorders [3], and inflammatory diseases [4], indicating the importance of stress responses in physiological and pathological processes.

Currently, diverse forms of physical stimuli are utilized as tools for various biological and clinical applications. Among these physical stimuli, pulsed electric fields (PEFs) have been proven particularly useful, because different biological effects can be achieved, depending on the duration of the electric pulses. PEFs in the range of milliseconds to microseconds primarily act on the cell membrane and generate membrane pores, which are suited for introduction of exogenous macromolecules, such as plasmid DNA, into living cells [5, 6]. Thus, these PEFs are widely used for DNA transfection [7].

Recent advances in electric engineering allow us to generate ultrashort PEFs in the range of nanoseconds, which are referred to as nanosecond PEFs (nsPEFs). Although nsPEFs do not generate membrane pores suitable for DNA transfection, they have been proven to be useful for cancer therapy, because of their ability to induce cell death [8, 9]. Furthermore, nsPEFs have been shown to induce stress responses that are mediated by phosphorylation of multiple translation initiation factors and eventually lead to transient suppression of general protein synthesis. For these reasons, nsPEFs have received considerable attention as a potential therapeutic method with a novel mechanism of action. This review is intended to provide an overview of the stress responses induced by nsPEFs. First, two major mechanisms for stress responses in human cells are explained. Phosphorylation-mediated control of two translation initiation factors, eIF2α and 4E-BP1, is critical in these pathways. Second, an outline of the biological actions of PEFs is provided, with particular emphasis on nsPEFs. Finally, stress responses induced by nsPEFs are described in detail.

2. Stress responses in human cells

2.1. Overview of stress response

The fundamental aspects of cellular stress responses are highly conserved among eukaryotes from yeast to humans. **Figure 1** shows a simplified scheme for eukaryotic stress responses. Under normal physiological conditions, cells continuously undergo protein synthesis, and the rate of protein synthesis is primarily regulated at the translation initiation step. When cells sense stress, they rapidly activate signal transduction that involves phosphorylation-mediated

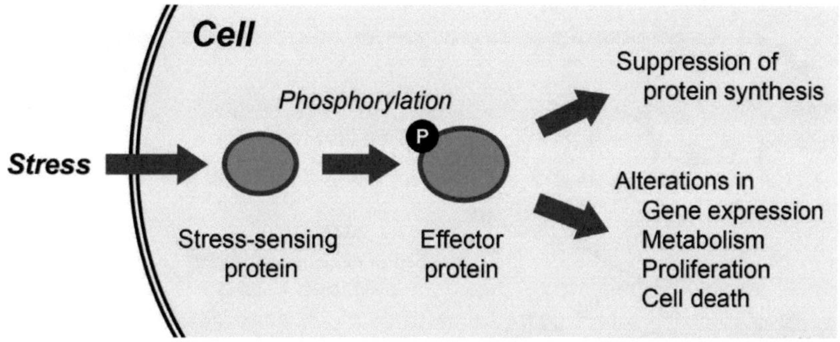

Figure 1. General principle of stress responses in eukaryotic cells. When a cell is exposed to stress, a stress-sensing protein is activated, inducing signal transduction mediated by protein phosphorylation. The stress-induced signal is transduced to downstream effector proteins. Stress responses generally lead to inhibition of cap-dependent translation initiation and consequent suppression of general protein synthesis. Stress responses often cause additional changes, such as alteration of gene expression.

control of translation initiation factors. Stress-induced alterations in the phosphorylation status of translation initiation factors reduce translation initiation and thereby result in attenuation of general protein synthesis. Because protein synthesis requires significant amounts of energy and materials, transient suppression of general protein synthesis conserves cellular resources and is thus beneficial for cells under stress. For these reasons, stress-induced suppression of general protein synthesis serves as a mechanism for survival. Once the stress ends, translation capacity is rapidly recovered by dephosphorylation of the translation factors. Intriguingly, persistent activation of stress responses is often associated with the induction of cell death, suggesting that timely induction and attenuation of stress responses are both critical for cell survival [10–12]. Although eukaryotic cells share a fundamental stress response mechanism [13, 14], as described below, human cells possess more intricate stress responses, including at least two distinct pathways involving phosphorylation of multiple proteins.

2.2. Stress response mediated by eIF2α phosphorylation

2.2.1. Stress-induced eIF2α phosphorylation and translational suppression

Translation initiation is a critical rate-limiting step in protein synthesis, and eukaryotic translation initiation factor 2 (eIF2) plays an essential role in this process [15]. eIF2 binds to guanine nucleotides, such as GDP and GTP. During initiation of translation, an eIF2-bound guanine nucleotide needs to cycle between GDP and GTP (see **Figure 2A** for details of GDP-GTP cycling in translation initiation). eIF2 consists of three subunits, α, β, and γ, and phosphorylation of the α subunit of eIF2 (eIF2α) is induced by various forms of stress [14, 16]. Phosphorylation of eIF2α at serine 51 interferes with GDP-GTP exchange on eIF2 [17] and consequently suppresses translation initiation (**Figure 2B**). The role of eIF2α phosphorylation in inhibition of translation initiation is highly conserved among eukaryotes, and the site of stress-induced phosphorylation (serine 51) is conserved in yeast and humans [13, 14].

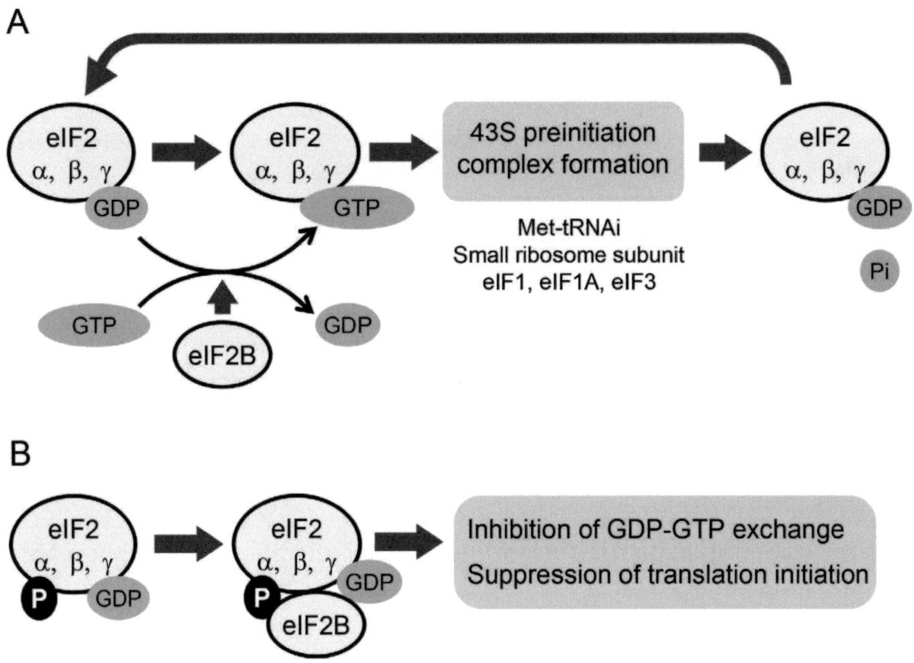

Figure 2. Suppression of translation initiation by eIF2α phosphorylation. (A) Role of GDP-GTP exchange on eIF2 in translation initiation. eIF2 is a trimer composed of three subunits, α, β, and γ, and binds to guanine nucleotides, such as GTP and GDP. For translation initiation, GTP-bound eIF2 recruits the initiator methionyl-tRNA (Met-tRNAi) and in turn forms the 43S preinitiation complex with the small ribosomal subunit and the initiation factors eIF1, eIF1A, and eIF3. The 43S preinitiation complex is recruited to the 5′ end of mRNA, which is marked with a cap structure, and scans the 5′ untranslated region of mRNA for the initiation codon. During this process, GTP on eIF2 is hydrolyzed to GDP, and GDP-bound eIF2 is released from the translation machinery. eIF2B, which has guanine exchange activity, replaces GDP on eIF2 with GTP, and GTP-bound eIF2 enters a new round of translation initiation [18, 19]. (B) Phosphorylation-mediated suppression of GDP-GTP exchange on eIF2. Under stressed conditions, serine 51 of eIF2α is rapidly phosphorylated by a stress-responsive protein kinase. Phosphorylation of eIF2α transforms eIF2 from a substrate into an inhibitor of eIF2B. eIF2B stalls on phosphorylated eIF2 and thus cannot exert its guanine exchange activity. Consequently, eIF2α phosphorylation leads to an increase in the GDP-bound form of eIF2, which is inactive for translation initiation, and results in attenuation of general protein synthesis [14, 16, 17].

Although initiation of translation of most mRNA species is dependent on their 5′ cap and is profoundly affected by eIF2α phosphorylation, a subset of mRNA species is actively translated under stressed conditions via alternative translation mechanisms that are independent of the cap structure. Approximately 2.5% of total mRNA is estimated to be preferentially translated under stressed conditions [20, 21], permitting synthesis of a subset of proteins that play critical roles in control of the stress response. For example, initiation of translation of ATF4 (activating transcription factor 4) mRNA is increased by ER stress caused by accumulation of unfolded proteins in the ER. The 5′ UTR of ATF4 mRNA contains two small open reading frames (upstream open reading frames [uORFs]), which play a critical role in initiation of translation under stressed conditions. ATF4 mRNA encodes a transcription factor that induces gene expression for adaptive responses. Thus, stress-induced eIF2α phosphorylation results in both suppression of general protein synthesis to conserve cellular resources and elevated translation of specific mRNA species for adaptive responses [22, 23].

http://dx.doi.org/10.5772/intechopen.69782

In human cells, phosphorylation of eIF2α is induced by a wide variety of exogenous as well as endogenous stresses, such as amino acid deprivation, UV irradiation, and accumulation of unfolded proteins in the ER. Although these stresses are sensed by individual mechanisms, multiple stress-induced events converge on a single reaction, namely, eIF2α phosphorylation. Thus, eIF2α phosphorylation integrates various stress-induced signals. The stress-induced signaling pathway in human and mammalian cells that involves eIF2α phosphorylation and downstream ATF4 induction is referred to as the integrated stress response (**Figure 3**) [12, 24].

2.2.2. *Stress-responsive protein kinases*

In human and mammalian cells, eIF2α is phosphorylated by four serine/threonine protein kinases, which are differentially activated by stress [14, 25]. These protein kinases show structural homology in their catalytic domains and are considered to exist in a monomeric form under unstressed conditions. Upon stress induction, they undergo homodimerization to become catalytically active, followed by autophosphorylation for full activation [26, 27].

PERK (protein kinase RNA-like endoplasmic reticulum kinase) is an ER transmembrane protein, and its N-terminal domain resides in the ER lumen and plays a role in sensing unfolded proteins. The C-terminal region of PERK is located in the cytoplasm and contains a kinase domain. PERK is activated by accumulation of unfolded proteins in the ER and in turn phosphorylates eIF2α.

GCN2 (general control nonderepressible 2) is critical for translational suppression under amino acid deprivation [28]. GCN2 binds to uncharged transfer RNAs and exerts its catalytic

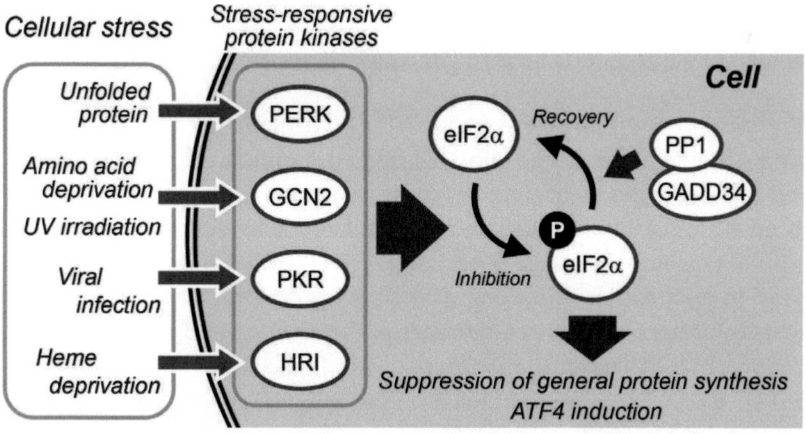

Figure 3. Integrated stress response. Eukaryotic cells respond to various stresses by inducing eIF2α phosphorylation that leads to suppression of general protein synthesis. Human and mammalian cells possess four stress-responsive protein kinases: PERK, GCN2, PKR, and HRI. These protein kinases are differentially activated by stress and in turn phosphorylate eIF2α. eIF2α phosphorylation and downstream ATF4 induction are known as the integrated stress responses, because different cellular reactions induced by various external cues converge to these reactions. eIF2α phosphorylation inhibits cap-dependent initiation of translation and consequently suppresses general protein synthesis. A complex of GADD34 and PP1 dephosphorylates eIF2α and serves as a negative feedback mechanism for the integrated stress response.

activity for eIF2α phosphorylation. In addition to amino acid deprivation, GCN2 has been reported to be activated by UV irradiation [29] and proteasome inhibition [30], although the activation mechanisms remain elusive.

PKR (double-stranded RNA-dependent protein kinase) was originally identified as a protein kinase activated by double-stranded RNA, which emerges during viral infection [31]. Phosphorylation of eIF2α by PKR interferes with translation of viral mRNA and thus serves as an antiviral mechanism. In addition, activation of PKR is involved in the pathology of obesity [32] and cancer [33], suggesting various physiological roles of PKR.

Heme-regulated inhibitor (HRI) has physiological roles particularly in erythroid tissues [34]. HRI is activated by heme deprivation and phosphorylates eIF2α to reduce globin synthesis under low-iron conditions [35].

2.2.3. Recovery from eIF2α phosphorylation-mediated translational suppression

Once stressed conditions end, eIF2α must be dephosphorylated to restore general protein synthesis. GADD34 (growth arrest and DNA damage-inducible protein 34) is a critical regulator of eIF2α dephosphorylation, and its activity in the relief of stress responses is controlled at both the transcriptional and translational levels. Expression of *GADD34* gene is low under normal physiological conditions and is activated by various forms of stress. GADD34 mRNA is translated by a cap-independent mechanism, in which uORFs in the 5′ UTR of GADD34 mRNA play critical roles [36]. GADD34 protein forms a complex with protein phosphatase 1 (PP1), yielding a catalytically active protein phosphatase that specifically catalyzes dephosphorylation of eIF2α to relieve translational suppression [37]. Thus, GADD34 constitutes a negative feedback mechanism for eIF2α-mediated translational suppression.

2.3. Stress response mediated by 4E-BP1 phosphorylation

Human cells have an alternative mechanism for stress-induced translational suppression, which involves 4E-binding protein 1 (4E-BP1) [38]. As mentioned above, most mRNA species are translated in a cap-dependent manner. eIF4E binds to the cap structure of mRNA and in turn recruits eIF4G and other translation initiation factors, resulting in formation of an active translation initiation complex on the 5′ end of mRNA. 4E-BP1 serves as a negative regulator of this process. In unstressed conditions, 4E-BP1 is highly phosphorylated at multiple sites, suppressing its inhibitory activity. Under energy deprivation and other stressed conditions, phosphorylation of 4E-BP1 is substantially decreased, and 4E-BP1 competes with eIF4G for binding to eIF4E, thereby inhibiting translation initiation (**Figure 4**) [38].

Phosphorylation of 4E-BP1 is primarily controlled by mTORC1 (mammalian target of rapamycin complex 1), which is a member of the phosphatidylinositol 3-kinase-related family of kinases [39]. Under normal physiological conditions, mTORC1 is catalytically active and suppresses the inhibitory activity of 4E-BP1 by phosphorylation. The kinase activity of mTORC1 is regulated by several cellular proteins, one of which is AMP-activated protein kinase (AMPK). AMPK senses the cellular energy status and negatively regulates mTORC1 activity [40]. AMPK, mTORC1, and 4E-BP1 constitute a stress-responsive mechanism for translational

http://dx.doi.org/10.5772/intechopen.69782

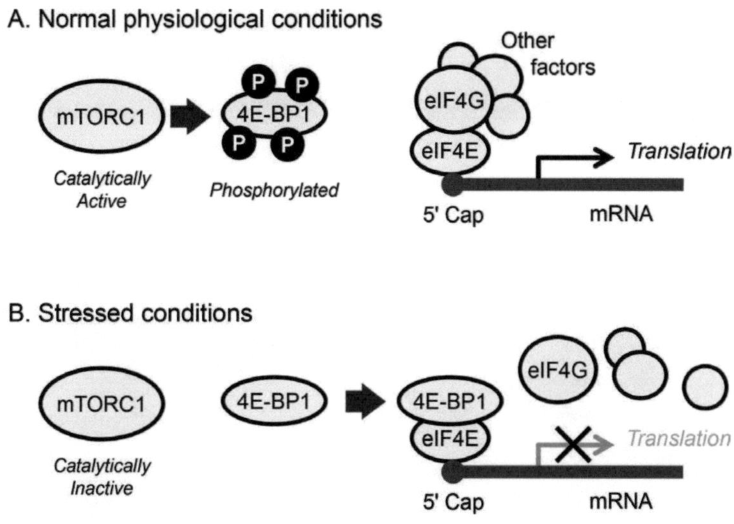

Figure 4. Control of translation initiation mediated by 4E-BP1 phosphorylation. Human cells possess an alternative mechanism for translational suppression, which is distinct from the eIF2α phosphorylation-mediated integrated stress response. eIF4E and other translation initiation factors form an active complex for initiation of translation at the cap structure of mRNA. 4E-BP1 serves as a negative regulator for the complex formation of eIF4E and other factors. (A) Under normal conditions, 4E-BP1 is highly phosphorylated by mTORC1 and sequestered from the translation initiation complex. (B) Under stressed conditions, such as energy deprivation, mTORC1 kinase activity is reduced, resulting in decreased 4E-BP1 phosphorylation, allowing inhibition of eIF4G-eIF4E binding.

suppression, independent of eIF2α phosphorylation. Thus, two mechanisms, the eIF2α-mediated integrated stress response and the 4E-BP1-mediated mechanism, function in translational suppression in human cells (**Figure 4**).

3. Pulsed electric fields as a novel physical tool in the life sciences

3.1. Effects of pulsed electric fields on living organisms

Pulsed electric fields (PEFs) refer to high-voltage electric pulses, which are milliseconds, microseconds, and nanoseconds in duration. PEFs have different effects on living organisms depending on pulse duration (**Figure 5**). PEFs with duration of milliseconds to microseconds primarily act on the cell membrane and cause pore formation. These membrane pores are suitable for transfer of macromolecules, such as plasmid DNA and drugs [5, 6]. Therefore, exposure of living cells to these PEFs is called electroporation and is commonly used for DNA transfection [7]. In addition, these PEFs are used for the introduction of antitumor drugs, which is called electrochemotherapy [41, 42]. Because PEFs with duration of milliseconds to microseconds primarily act on the cell membrane, these PEFs often induce cell death via cell membrane damage.

Recent advances in electrical engineering have enabled the generation of high-voltage electric pulses for ultrashort periods in the nanosecond range, which are called nanosecond PEFs

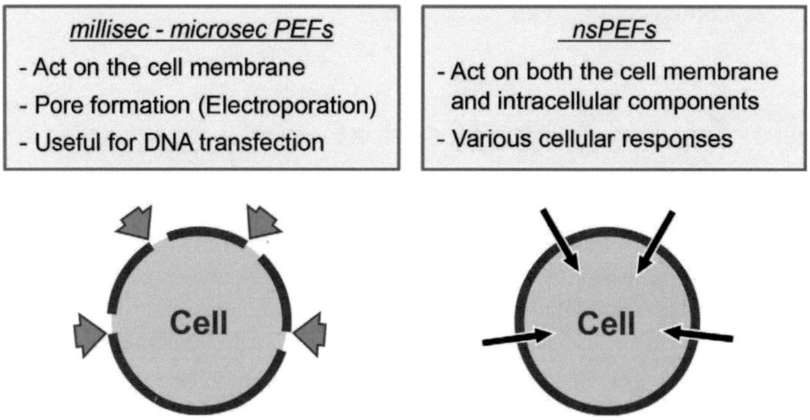

Figure 5. Comparison between electroporation and nsPEF action. PEFs have different effects on living organisms depending on pulse duration. (*Left*) PEFs with duration of milliseconds to microseconds are widely used for electroporation, because these PEFs primarily act on the cell membrane and generate membrane pores suited for macromolecule transfer. (*Right*) nsPEFs generate small membrane pores that permeate small molecules, such as ions and water. Furthermore, nsPEFs have been suggested to directly affect intracellular components. Although nsPEFs are unsuitable for DNA transfection and cancer electrochemotherapy by electroporation, these PEFs can induce various cellular responses, including cell death induction.

(nsPEFs). It has become increasingly evident that nsPEFs have unique biological actions distinct from electroporation. The pulse duration of nsPEFs is too short to generate membrane pores large enough for entry of macromolecules. Thus, nsPEFs are generally unsuitable for DNA transfection and cancer electrochemotherapy. Instead, nsPEFs produce very small membrane pores that allow passage of small molecules, such as ions and water [43–45]. Accordingly, exposure of cultured human cells to nsPEFs causes Ca^{2+} influx and membrane blebbing due to ion imbalance across the cell membrane [46–48].

Whereas millisecond-to-microsecond PEFs primarily act on the cell membrane, previous theoretical studies have strongly suggested that nsPEFs exert their effects on both the cell membrane and intracellular components [49, 50]. In accordance, extensive biochemical analyses have proven that nsPEFs elicit various intracellular responses, as described below.

3.2. Cellular responses to nsPEFs

Recent studies have revealed that nsPEFs elicit different intracellular responses in a manner dependent on nsPEF intensity. **Figure 6** summarizes the relationship between nsPEF intensity and intracellular responses in human cells. Relatively weak nsPEFs do not cause morphological changes observable under a microscope, growth retardation, or cell death. However, cells rapidly respond to such stimuli by activating multiple intracellular signal pathways, including MAPK pathways [51, 52] and AMPK pathway [53]. Intracellular signaling is mediated by sequential phosphorylation of proteins in these pathways, leading to expression of downstream genes [51, 52]. When moderate-intensity nsPEFs are used, two independent mechanisms for stress responses are activated [54]. Intense nsPEFs efficiently induce cell death in vitro [8] as well as in vivo [9].

http://dx.doi.org/10.5772/intechopen.69782

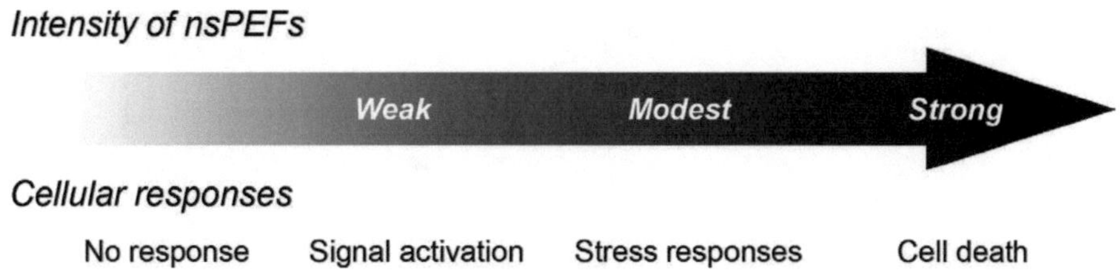

Figure 6. Relationship between nsPEF intensity and cellular responses. Exposure of human cells to nsPEFs causes different cellular responses, depending on intensity of nsPEFs. Relatively weak nsPEFs activate several signal transduction pathways, such as MAPK pathways, and their downstream gene expression without affecting on cell viability. Moderate-intensity nsPEFs elicit stress responses and cause growth retardation. Intense nsPEFs induce either apoptotic or necrotic cell death in a cell-type-dependent manner and can be used for cancer therapy.

When cultured human cells are exposed to intense nsPEFs, cell-type dependency of cell death modes has been observed. For example, apoptosis is induced in HL-60 and Jurkat cells by intense nsPEFs [8, 55, 56], whereas necrotic cell death is elicited in several cell lines, including U937, K562, and HeLa S3 [56, 57], demonstrating that the cellular context determines the mode of cell death. Induction of necrosis by intense nsPEFs is a Ca^{2+}-dependent process [48, 58], while nsPEF-induced apoptosis is largely unaffected by the presence or absence of Ca^{2+} [58].

Because of the ability of nsPEFs to induce efficient cell death, many studies have attempted to apply nsPEFs to cancer therapy and have demonstrated their effectiveness in several mouse models [9, 59, 60]. Furthermore, a human clinical trial of nsPEFs for cancer therapy has started [61]. For these reasons, nsPEFs have attracted much interest and are increasingly recognized as a novel method of cancer therapy.

4. Induction of stress responses by nsPEFs

Previously, nsPEFs have been shown to activate two independent stress-responsive pathways, both of which are mediated by protein phosphorylation and result in translational suppression. First, nsPEFs induce eIF2α phosphorylation via two protein kinases, PERK and GCN2. Concomitant with eIF2α phosphorylation, GADD34 expression is elevated. Second, nsPEFs cause decreased phosphorylation of 4E-BP1, which is presumably controlled by mTORC1. Following induction of these phosphorylation-mediated stress responses, general protein synthesis is markedly reduced in nsPEF-exposed cells. **Figure 7** represents a summary of nsPEF-induced stress responses.

4.1. eIF2α phosphorylation induced by nsPEFs

Most mRNA species are translated in a cap-dependent manner, and eIF2 plays a critical role in this process. Under stressed conditions, the activity of eIF2 for translation initiation is suppressed by phosphorylation of its α subunit (eIF2α) [14]. eIF2α phosphorylation is considered to be a hallmark of induction of the integrated stress response and can be examined by

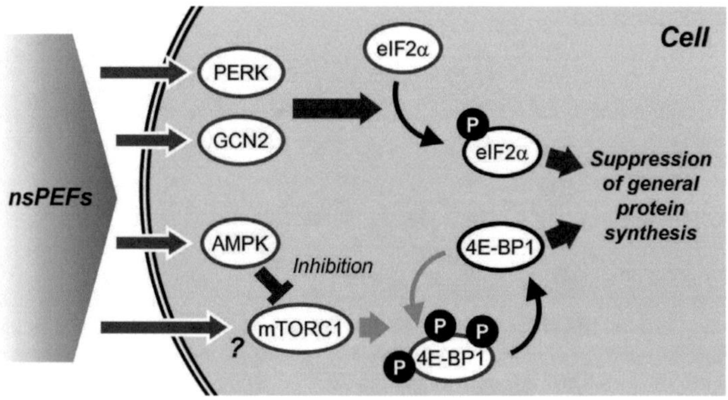

Figure 7. Stress responses induced by nsPEFs. nsPEFs affect multiple stress-responsive proteins in the two independent signaling pathways, both of which result in suppression of general protein synthesis. The first signaling pathway induced by nsPEFs is mediated by eIF2α phosphorylation. nsPEFs activate PERK and GCN2, which in turn phosphorylate eIF2α. eIF2α phosphorylation interferes GDP-GTP exchange in translation initiation, resulting in suppression of general protein synthesis. The second pathway induced by nsPEFs involves a reduction in 4E-BP1 phosphorylation. 4E-BP1 is highly phosphorylated by mTORC1 under normal conditions. nsPEFs cause a decrease in 4E-BP1 phosphorylation, leading to inhibition of eIF4G-eIF4E complex formation and consequent suppression of translation initiation. nsPEFs are known to activate AMPK, which is known to function as a negative regulator of mTORC1 under energy deprivation conditions. AMPK activation may account for the decrease in 4E-BP1 phosphorylation following nsPEF exposure. Direct effects of nsPEFs on mTORC1 will be examined by future research. Increased eIF2α phosphorylation and decreased 4E-BP1 phosphorylation serve as two independent mechanisms for the suppression of general protein synthesis.

Western blotting using an antibody specific to phosphorylated eIF2α. When cultured cells are exposed to nsPEFs, eIF2α phosphorylation can be detected, indicating that nsPEFs activate the integrated stress response [54]. nsPEF-induced eIF2α phosphorylation can be detected in all cell lines examined so far, which include HeLa S3, HCT116, Jurkat, and mouse embryonic fibroblasts (MEFs). In HeLa S3 cells, eIF2α phosphorylation is detectable within 1 min after nsPEF exposure, persists at high levels for 30 min, and decreases thereafter [54]. GADD34 is known to play a critical role in recovery from the integrated stress response. Under stressed conditions, GADD34 is positively controlled at the transcriptional and translational levels. Consistently, expression of GADD34 is significantly activated in nsPEF-exposed cells, suggesting that GADD34 is involved in recovery from the nsPEF-induced stress response [54].

As described above, cellular responses to nsPEFs are dependent on nsPEF intensity (**Figure 6**). Relatively mild nsPEFs activate several signal transduction pathways, such as MAPK pathways, but are insufficient to induce eIF2α phosphorylation. Moderate levels of nsPEF intensity are required for induction of eIF2α phosphorylation. Such nsPEFs also cause retardation in cell proliferation but not cell death [54]. Intense nsPEFs induce eIF2α phosphorylation and cell death. Currently, it remains unknown whether the nsPEF-induced stress response positively affects cell survival or facilitates cell death induction.

4.2. Participation of PERK and GCN2 in nsPEF-induced eIF2α phosphorylation

Human and mammalian cells have four stress-responsive protein kinases for eIF2α phosphorylation. These kinases differentially respond to various forms of stress, and at least one of them is activated for eIF2α phosphorylation [14]. Autophosphorylation is critical for the activation of these kinases and can be analyzed by Western blot analysis using antibodies against phosphorylated forms of these kinases. In nsPEF-exposed cells, PERK and GCN2 are activated, as shown by their autophosphorylation [54]. Experiments using PERK and GCN2 knockout cells suggest that these kinases play mutually compensatory roles in nsPEF-induced eIF2α phosphorylation. MEFs lacking either *PERK* or *GCN2* gene display nsPEF-induced eIF2α phosphorylation comparable to that in wild-type cells. However, double-knockout cells lacking both *PERK* and *GCN2* genes exhibit a significant reduction in nsPEF-induced eIF2α phosphorylation [54]. These observations suggest that PERK and GCN2 perform redundant functions in nsPEF-induced eIF2α phosphorylation.

Because PERK is well known to be activated by ER stress, the observation on the nsPEF-induced PERK activation raised the possibility that nsPEFs cause ER stress. To clarify this point, downstream events in the ER stress pathway were analyzed [54]. The signal pathway induced by ER stress generally leads to alterations of gene expression [62]. Transcription of *CHOP* (CCAAT-enhancer-binding protein homologous protein) gene is a major downstream event in the ER stress response. In addition, mRNA for XBP1 (X-box-binding protein 1) is known to undergo alternative splicing after ER stress induction [63]. However, quantitative RT-PCR analysis of *CHOP* and *XBP1* mRNAs demonstrated that nsPEF-exposed cells showed neither elevated CHOP expression nor altered XBP1 splicing. Furthermore, UV irradiation has been reported to induce activation of GCN2 and transcription of downstream genes, such as *GADD45* (growth arrest and DNA damage-inducible 45) [64, 65], but quantitative RT-PCR analysis showed no substantial changes in these mRNAs in nsPEF-exposed cells. These observations indicate that nsPEFs exert their effects in a manner that is distinct from ER stress or UV irradiation, although nsPEFs activate PERK and GCN2 [54].

4.3. Decreased 4E-BP1 phosphorylation by nsPEFs

In addition to the eIF2α-mediated response, a distinct mechanism involving 4E-BP1 phosphorylation is known to play a critical role in stress-induced translational suppression [38]. Under normal physiological conditions, 4E-BP1 is highly phosphorylated by mTORC1 [39], and hyperphosphorylation suppresses its inhibition of cap-dependent translation initiation. Stressed conditions, particularly energy deprivation, reduce mTORC1 activity and result in decreased 4E-BP1 phosphorylation. 4E-BP1 at low phosphorylation status interferes with assembly of translation factors on the cap structure of mRNAs and thereby suppresses general protein synthesis (**Figure 4**). In nsPEF-exposed cells, 4E-BP1 phosphorylation is rapidly decreased [54], suggesting that nsPEFs activate stress responses mediated by 4E-BP1 phosphorylation. The decrease in 4E-BP1 phosphorylation is indistinguishable between wild-type cells and *PERK/GCN2* double-knockout cells, supporting the idea that nsPEFs activate two independent mechanisms.

The decrease in 4E-BP1 phosphorylation following nsPEF exposure suggests that nsPEFs cause a reduction in the catalytic activity of mTORC1. A previous study demonstrated that AMPK is rapidly activated by nsPEFs [53]. AMPK functions as an energy sensor and is activated by elevated intracellular AMP levels, which are primarily caused by energy deprivation

[40]. The catalytic activity of mTORC1 is well known to be negatively regulated by AMPK [39]. A previous study has demonstrated that, concomitant with AMPK activation, nsPEFs induce phosphorylation of AMPK substrates, such as acetyl-CoA carboxylase-2 [53], suggesting that nsPEF-activated AMPK also phosphorylates other substrates, including mTORC1. Although mTORC1 catalytic activity has not been examined in nsPEF-exposed cells yet, the above observations suggest that AMPK downregulates mTORC1, leading to reduced 4E-BP1 phosphorylation in nsPEF-exposed cells (**Figure 7**).

4.4. Suppression of general protein synthesis by nsPEFs

As described above, nsPEFs cause elevated eIF2α phosphorylation and decrease 4E-BP1 phosphorylation, both of which are known to be involved in suppression of general protein synthesis. To test whether exposure to nsPEFs actually leads to translational suppression, measurement of protein synthesis rates was required. To this end, metabolic labeling of newly synthesized proteins with radioactive amino acids was employed [54]. **Figure 8** shows an outline of metabolic labeling using ^{35}S-labeled amino acids. Using this method, suppression of general protein synthesis in nsPEF-exposed cells was demonstrated [54]. After nsPEF exposure, overall protein synthesis quickly decreased, and maximum suppression of protein synthesis was observed at 30 min. Protein synthesis in nsPEF-exposed cells recovered to approximately 80% within 2 h. When cells were treated with UV irradiation, general protein synthesis decreased gradually for several hours. Compared to UV irradiation, nsPEFs cause acute translational suppression, and recovery is more rapid than in UV-irradiated cells.

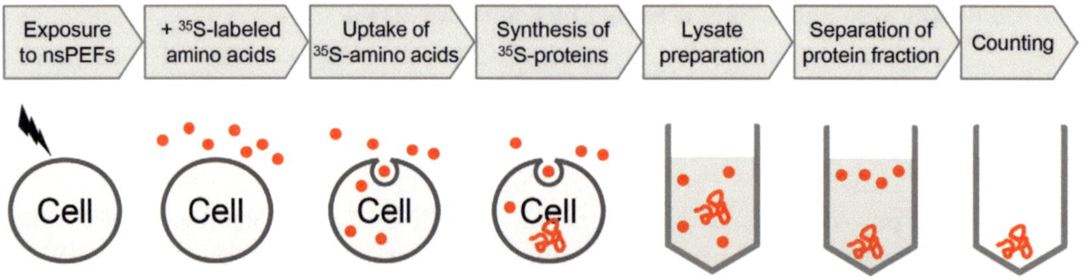

Figure 8. Measurement of protein synthesis rates by metabolic labeling of newly synthesized proteins with radioactive amino acids. A rate of protein synthesis can be measured as incorporation of radioactive amino acids into cellular proteins. Following appropriate treatment, such as nsPEF exposure, cells are incubated in culture medium containing ^{35}S-labeled methionine and cysteine. During incubation, cells use radioactive amino acids to synthesize proteins, yielding ^{35}S-labeled proteins. Following preparation of whole-cell lysate, the protein fraction is separated from the free amino acids, and the radioactivity incorporated into the proteins is quantified by liquid scintillation counting.

5. Conclusion

Exposure of cultured human cells to nsPEFs elicits two distinct stress responses, both of which are controlled by phosphorylation of translation initiation factors. nsPEFs rapidly induce eIF2α phosphorylation and concomitant activation of the stress-responsive kinases,

PERK and GCN2. In addition, nsPEFs cause decreased 4E-BP1 phosphorylation and AMPK activation, which appear to constitute a stress response pathway involving mTORC1. nsPEFs elicit acute suppression of general protein synthesis via two reactions for inhibition of translation initiation. Collectively, these findings clearly indicate that nsPEFs act as a novel form of cellular stress and suppress general protein synthesis.

Although the identification of key events in nsPEF-induced stress responses has significantly advanced our understanding of the biological effects of nsPEFs, several critical questions remain to be elucidated. First, the site of action of nsPEFs for eIF2α-mediated stress response is currently unclear. Because PERK and GCN2 are the most upstream molecules in their signaling pathways, nsPEFs may act directly on these kinases, causing eIF2α phosphorylation. Second, the decrease in 4E-BP1 phosphorylation and the activation of AMPK strongly suggest that mTORC1 participates in the nsPEF-induced stress response, because energy deprivation sequentially causes AMPK activation, reduced mTORC1 activity, and consequent decreased phosphorylation of 4E-BP1. To test this idea, the relationships among AMPK, mTORC1, and 4E-BP1 in nsPEF-exposed cells should be investigated in detail. Furthermore, analysis of effects of nsPEFs on cellular energy levels is also important for understanding how nsPEFs control 4E-BP1 phosphorylation.

Finally, the significance of nsPEF-induced stress responses for cell survival should be determined. Suppression of general protein synthesis under stressed conditions conserves biological resources and is regarded as an important mechanism for cell survival [2]. However, prolonged activation of stress responses often has an opposite effect, facilitating the induction of cell death, presumably because elimination of overstressed cells is beneficial for the body [10, 14]. Currently, it remains unclear whether nsPEF-induced stress responses serve as a prosurvival mechanism or serve to facilitate cell death induction. Future efforts will focus on understanding the contribution of nsPEF-induced stress responses to cell survival and death. Previous studies have revealed the unique effects of nsPEFs as a novel form of cellular stress. More detailed understanding of the molecular mechanisms and biological importance of nsPEF-induced stress responses will pave the way toward more effective applications of this novel technology in a wide range of biomedical sciences.

Acknowledgements

This work was supported by JSPS KAKENHI grant numbers 26350540 and 16H02311 and by a grant from the SEI Group CSR Foundation.

Author details

Ken-ichi Yano* and Keiko Morotomi-Yano

*Address all correspondence to: yanoken@kumamoto-u.ac.jp

Institute of Pulsed Power Science, Kumamoto University, Kumamoto, Japan

References

[1] Alberts B, Johnson A, Lewis J, Morgan D, Raff M, Roberts K and Walter P. Molecular Biology of the Cell, 6th ed. Garland Publishing, Inc; New York, NY, USA; 2014

[2] Leprivier G, Rotblat B, Khan D, Jan E and Sorensen PH. Stress-mediated translational control in cancer cells. Biochimica et Biophysica Acta. 2015;**1849**:845-860

[3] Duennwald ML. Cellular stress responses in protein misfolding diseases. Future Science OA. 2015;**1**:FSO42

[4] Grootjans J, Kaser A, Kaufman RJ and Blumberg RS. The unfolded protein response in immunity and inflammation. Nature Reviews Immunology. 2016;**16**:469-484

[5] Rems L and Miklavcic D. Tutorial: Electroporation of cells in complex materials and tissue. Journal of Applied Physics. 2016;**119**:201101

[6] Teissie J. Electropermeabilization of the cell membrane. Methods in Molecular Biology. 2014;**1121**:25-46

[7] Heller R and Heller LC. Gene electrotransfer clinical trials. Advances in Genetics. 2015;**89**:235-262

[8] Beebe SJ, Fox PM, Rec LJ, Willis EL and Schoenbach KH. Nanosecond, high-intensity pulsed electric fields induce apoptosis in human cells. FASEB Journal. 2003;**17**:1493-1495

[9] Nuccitelli R, Chen X, Pakhomov AG, Baldwin WH, Sheikh S, Pomicter JL, Ren W, Osgood C, Swanson RJ, Kolb JF et al. A new pulsed electric field therapy for melanoma disrupts the tumor's blood supply and causes complete remission without recurrence. International Journal of Cancer. 2009;**125**:438-445

[10] Woehlbier U and Hetz C. Modulating stress responses by the UPRosome: A matter of life and death. Trends in Biochemical Sciences. 2011;**36**:329-337

[11] Sano R and Reed JC. ER stress-induced cell death mechanisms. Biochimica et Biophysica Acta. 2013;**1833**:3460-3470

[12] Pakos-Zebrucka K, Koryga I, Mnich K, Ljujic M, Samali A and Gorman AM. The integrated stress response. EMBO Reports. 2016;**17**:1374-1395

[13] Hinnebusch AG. Translational regulation of GCN4 and the general amino acid control of yeast. Annual Review of Microbiology. 2005;**59**:407-450

[14] Koromilas AE. Roles of the translation initiation factor eIF2alpha serine 51 phosphorylation in cancer formation and treatment. Biochimica et Biophysica Acta. 2015;**1849**:871-880

[15] Proud CG. eIF2 and the control of cell physiology. Seminars in Cell and Developmental Biology. 2005;**16**:3-12

[16] Spriggs KA, Bushell M and Willis AE. Translational regulation of gene expression during conditions of cell stress. Molecular Cell. 2010;**40**:228-237

[17] Rowlands AG, Panniers R and Henshaw EC. The catalytic mechanism of guanine nucleotide exchange factor action and competitive inhibition by phosphorylated eukaryotic initiation factor 2. Journal of Biological Chemistry. 1988;**263**:5526-5533

[18] Sonenberg N and Hinnebusch AG. Regulation of translation initiation in eukaryotes: Mechanisms and biological targets. Cell. 2009;**136**:731-745

[19] Hinnebusch AG. The scanning mechanism of eukaryotic translation initiation. Annual Review of Biochemistry. 2014;**83**:779-812

[20] Dang Do AN, Kimball SR, Cavener DR and Jefferson LS. eIF2alpha kinases GCN2 and PERK modulate transcription and translation of distinct sets of mRNAs in mouse liver. Physiological Genomics. 2009;**38**:328-341

[21] Baird TD and Wek RC. Eukaryotic initiation factor 2 phosphorylation and translational control in metabolism. Advances in Nutrition. 2012;**3**:307-321

[22] Harding HP, Zhang Y, Zeng H, Novoa I, Lu PD, Calfon M, Sadri N, Yun C, Popko B, Paules R, et al. An integrated stress response regulates amino acid metabolism and resistance to oxidative stress. Molecular Cell. 2003;**11**:619-633

[23] Vattem KM and Wek RC. Reinitiation involving upstream ORFs regulates ATF4 mRNA translation in mammalian cells. Proceedings of the National Academy of Sciences of the United States of America. 2004;**101**:11269-11274

[24] Ron D. Translational control in the endoplasmic reticulum stress response. Journal of Clinical Investigation. 2002;**110**:1383-1388

[25] Donnelly N, Gorman AM, Gupta S and Samali A. The eIF2alpha kinases: Their structures and functions. Cellular and Molecular Life Sciences. 2013;**70**:3493-3511

[26] Dey M, Cao C, Dar AC, Tamura T, Ozato K, Sicheri F and Dever TE. Mechanistic link between PKR dimerization, autophosphorylation, and eIF2alpha substrate recognition. Cell. 2005;**122**:901-913

[27] Lavoie H, Li JJ, Thevakumaran N, Therrien M and Sicheri F. Dimerization-induced allostery in protein kinase regulation. Trends in Biochemical Sciences. 2014;**39**:475-486

[28] Zhang P, McGrath BC, Reinert J, Olsen DS, Lei L, Gill S, Wek SA, Vattem KM, Wek RC, Kimball SR, et al. The GCN2 eIF2alpha kinase is required for adaptation to amino acid deprivation in mice. Molecular Cell. Biology. 2002;**22**:6681-6688

[29] Deng J, Harding HP, Raught B, Gingras AC, Berlanga JJ, Scheuner D, Kaufman RJ, Ron D and Sonenberg N. Activation of GCN2 in UV-irradiated cells inhibits translation. Current Biology. 2002;**12**:1279-1286

[30] Jiang HY and Wek RC. Phosphorylation of the alpha-subunit of the eukaryotic initiation factor-2 (eIF2alpha) reduces protein synthesis and enhances apoptosis in response to proteasome inhibition. Journal of Biological Chemistry. 2005;**280**:14189-14202

[31] Meurs E, Chong K, Galabru J, Thomas NS, Kerr IM, Williams BR and Hovanessian AG. Molecular cloning and characterization of the human double-stranded RNA-activated protein kinase induced by interferon. Cell. 1990;**62**:379-390

[32] Nakamura T, Furuhashi M, Li P, Cao H, Tuncman G, Sonenberg N, Gorgun CZ and Hotamisligil GS. Double-stranded RNA-dependent protein kinase links pathogen sensing with stress and metabolic homeostasis. Cell. 2010;**140**:338-348

[33] Kim SH, Gunnery S, Choe JK and Mathews MB. Neoplastic progression in melanoma and colon cancer is associated with increased expression and activity of the interferon-inducible protein kinase, PKR. Oncogene. 2002;**21**:8741-8748

[34] Crosby JS, Lee K, London IM and Chen JJ. Erythroid expression of the heme-regulated eIF-2 alpha kinase. Molecular and Cellular Biology. 1994;**14**:3906-3914

[35] Han AP, Yu C, Lu L, Fujiwara Y, Browne C, Chin G, Fleming M, Leboulch P, Orkin SH and Chen JJ. Heme-regulated eIF2alpha kinase (HRI) is required for translational regulation and survival of erythroid precursors in iron deficiency. EMBO Journal. 2001;**20**:6909-6918

[36] Lee YY, Cevallos RC and Jan E. An upstream open reading frame regulates translation of GADD34 during cellular stresses that induce eIF2alpha phosphorylation. Journal of Biological Chemistry. 2009;**284**:6661-6673

[37] Novoa I, Zeng H, Harding HP and Ron D. Feedback inhibition of the unfolded protein response by GADD34-mediated dephosphorylation of eIF2alpha. Journal of Cell Biology. 2001;**153**:1011-1022

[38] Ma XM and Blenis J. Molecular mechanisms of mTOR-mediated translational control. Nature Reviews Molecular Cell Biology. 2009;**10**:307-318

[39] Heberle AM, Prentzell MT, van Eunen K, Bakker BM, Grellscheid SN and Thedieck K. Molecular mechanisms of mTOR regulation by stress. Molecular and Cellular Oncology. 2015;**2**:e970489

[40] Hardie DG and Ashford ML. AMPK: Regulating energy balance at the cellular and whole body levels. Physiology (Bethesda). 2014;**29**:99-107

[41] Mir LM. Electroporation-based gene therapy: Recent evolution in the mechanism description and technology developments. Methods in Molecular Biology. 2014;**1121**:3-23

[42] Miklavcic D, Mali B, Kos B, Heller R and Sersa G. Electrochemotherapy: From the drawing board into medical practice. Biomedical Engineering Online. 2014;**13**:29

[43] Vernier PT, Sun Y and Gundersen MA. Nanoelectropulse-driven membrane perturbation and small molecule permeabilization. BMC Cell Biology. 2006;**7**:37

[44] Pakhomov AG, Kolb JF, White JA, Joshi RP, Xiao S and Schoenbach KH. Long-lasting plasma membrane permeabilization in mammalian cells by nanosecond pulsed electric field (nsPEF). Bioelectromagnetics. 2007;**28**:655-663

[45] Silve A, Leray I and Mir LM. Demonstration of cell membrane permeabilization to medium-sized molecules caused by a single 10 ns electric pulse. Bioelectrochemistry. 2012;**87**:260-264

[46] Vernier PT, Sun Y, Marcu L, Salemi S, Craft CM and Gundersen MA. Calcium bursts induced by nanosecond electric pulses. Biochemical and Biophysical Research Communications. 2003;**310**:286-295

[47] White JA, Blackmore PF, Schoenbach KH and Beebe SJ. Stimulation of capacitative calcium entry in HL-60 cells by nanosecond pulsed electric fields. Journal of Biological Chemistry. 2004;**279**:22964-22972

[48] Pakhomova ON, Gregory B, Semenov I and Pakhomov AG. Calcium-mediated pore expansion and cell death following nanoelectroporation. Biochimica et Biophysica Acta. 2014;**1838**:2547-2554

[49] Schoenbach KH, Beebe SJ and Buescher ES. Intracellular effect of ultrashort electrical pulses. Bioelectromagnetics. 2001;**22**:440-448

[50] Beebe SJ and Schoenbach KH. Nanosecond pulsed electric fields: A new stimulus to activate intracellular signaling. Journal of Biomedicine and Biotechnology. 2005;**2005**:297-300

[51] Morotomi-Yano K, Uemura Y, Katsuki S, Akiyama H and Yano K. Activation of the JNK pathway by nanosecond pulsed electric fields. Biochemical and Biophysical Research Communications. 2011;**408**:471-476

[52] Morotomi-Yano K, Akiyama H and Yano K. Nanosecond pulsed electric fields activate MAPK pathways in human cells. Archives of Biochemistry and Biophysics. 2011;**515**:99-106

[53] Morotomi-Yano K, Akiyama H and Yano K. Nanosecond pulsed electric fields activate AMP-activated protein kinase: Implications for calcium-mediated activation of cellular signaling. Biochemical and Biophysical Research Communications. 2012;**428**:371-375

[54] Morotomi-Yano K, Oyadomari S, Akiyama H and Yano K. Nanosecond pulsed electric fields act as a novel cellular stress that induces translational suppression accompanied by eIF2alpha phosphorylation and 4E-BP1 dephosphorylation. Experimental Cell Research. 2012;**318**:1733-1744

[55] Ren W, Sain NM and Beebe SJ. Nanosecond pulsed electric fields (nsPEFs) activate intrinsic caspase-dependent and caspase-independent cell death in Jurkat cells. Biochemical and Biophysical Research Communications. 2012;**421**:808-812

[56] Morotomi-Yano K, Akiyama H and Yano K. Nanosecond pulsed electric fields induce poly(ADP-ribose) formation and non-apoptotic cell death in HeLa S3 cells. Biochemical and Biophysical Research Communications. 2013;**438**:557-562

[57] Pakhomova ON, Gregory BW, Semenov I and Pakhomov AG. Two modes of cell death caused by exposure to nanosecond pulsed electric field. PLoS One. 2013;**8**:e70278

[58] Morotomi-Yano K, Akiyama H and Yano K. Different involvement of extracellular calcium in two modes of cell death induced by nanosecond pulsed electric fields. Archives of Biochemistry and Biophysics. 2014;**555-556**:47-54

[59] Nuccitelli R, Tran K, Athos B, Kreis M, Nuccitelli P, Chang KS, Epstein EH, Jr. and Tang JY. Nanoelectroablation therapy for murine basal cell carcinoma. Biochemical and Biophysical Research Communications. 2012;**424**:446-450

[60] Nuccitelli R, Huynh J, Lui K, Wood R, Kreis M, Athos B and Nuccitelli P. Nanoelectroablation of human pancreatic carcinoma in a murine xenograft model without recurrence. International Journal of Cancer. 2013;**132**:1933-1939

[61] Nuccitelli R, Wood R, Kreis M, Athos B, Huynh J, Lui K, Nuccitelli P and Epstein EH, Jr. First-in-human trial of nanoelectroablation therapy for basal cell carcinoma: Proof of method. Experimental Dermatology. 2014;**23**:135-137

[62] Szegezdi E, Logue SE, Gorman AM and Samali A. Mediators of endoplasmic reticulum stress-induced apoptosis. EMBO Reports. 2006;**7**:880-885

[63] Yoshida H, Matsui T, Yamamoto A, Okada T and Mori K. XBP1 mRNA is induced by ATF6 and spliced by IRE1 in response to ER stress to produce a highly active transcription factor. Cell. 2001;**107**:881-891

[64] Takekawa M and Saito H. A family of stress-inducible GADD45-like proteins mediate activation of the stress-responsive MTK1/MEKK4 MAPKKK. Cell. 1998;**95**:521-530

[65] Campbell KJ, Rocha S and Perkins ND. Active repression of antiapoptotic gene expression by RelA(p65) NF-kappa B. Molecular Cell. 2004;**13**:853-865

Modification of α-Synuclein by Phosphorylation: A Pivotal Event in the Cellular Pathogenesis of Parkinson's Disease

Indrani Datta and Kavina Ganapathy

Additional information is available at the end of the chapter

http://dx.doi.org/10.5772/intechopen.70405

Abstract

Post-translational protein modifications play an important role in generating the large diversity of the proteome in comparison to the relatively small number of genes; phosphorylation being the most widespread. Phosphorylation of proteins regulates important molecular-switches for several cellular events and abnormal phosphorylation events are associated in many neurodegenerative diseases. In Parkinson's disease (PD) the main hallmark is the accumulation of cytoplasmic inclusions, Lewy bodies (LBs), consisting of α-synuclein (α-Syn) and ubiquitin. There's another key observation which is increasingly gaining prominence is a modified-form of α-Syn; the phospho α-Syn serine129 (pSyn). The significance of pSyn has gained importance in PD because its accumulation is distinctly enhanced in the diseased condition. The revelation of the involvement of pSyn on α-Syn aggregation, LB formation and neurotoxicity is crucial to understanding the pathogenesis and progression of PD. Since some *in vitro* and *in vivo* studies have indicated that pSyn is an early event preceding apoptosis, some important questions now needs to be explored in reference to the physiological functions regulated by phosphorylation, such as dopamine synthesis, vesicle mobilization, regulation of synaptic proteins, and synaptic plasticity. An investigation of the role of enzymes on the phosphorylation and clearance of α-Syn and region-specific susceptibility is required to be determined; to identify viable targets for new therapeutics.

Keywords: α-synuclein, phospho α-synuclein serine129, PD, phosphorylation, kinases, aggregation, biomarker, neurotoxicity, Lewy bodies

1. Introduction

Parkinson's disease (PD) is the second most common diagnosed neurodegenerative disease [1] with a prevalence of about 1% at the age of 65 and of 4–5% by the age of 85 [2]. The clinical manifestations of classical PD are rest tremor, rigidity, bradykinesia, and postural imbalance. The loss of dopaminergic (DA) neurons in the substantia nigra pars compacta (SNpc) together with a distinct decrease in striatal dopamine, and the occurrence of cytoplasmic eosinophilic inclusions called Lewy bodies (LBs) are considered the pathological hallmarks of PD [3]. There are also reports of other neuronal cell losses in locus coeruleus and olfactory lobe during the development of the disease [4]. It has however been identified that the clinical manifestation of motor symptoms appears with the loss of DA neurons in the midbrain [5]. According to Rodriguez-Oroz et al. [6], the anatomical-functional basis of the main clinical manifestations is related to the low level of dopamine concentration in contralateral striatum and the malfunction of dopamine circuits. Current medications for PD supplement dopamine (L-DOPA), or activate DA receptors (DA-receptor agonist), or inhibit the degradation of DA (monoamine oxidase B inhibitor and catechol-O-methyltransferase inhibitor), bringing about temporary abatement of motor symptoms but failing to delay or halt disease progression. The etiology of PD is still unknown: it comprises familial (fPD) forms accounting for less than 10% of all PD cases, and the far more common sporadic (sPD) form. The striking feature in both fPD and sPD is α-synuclein (α-Syn) aggregation with ubiquitin that eventually progresses to form LBs. Three missense mutations (A53T, A30P, and E46K) [7–9] of α-Syn are known to cause autosomal dominant fPD [10], probably through a gain-of-function mechanism. Moreover, overexpression of human α-Syn in mice results in progressive loss of DA terminals in the basal ganglia and accumulation of LB-like structures in neurons [11]. Mechanisms that might control α-Syn aggregation in sPD are not clear, but may include transcription factor dysregulation [12] and the inability of normal degradation pathways to function adequately [13]. El-Agnaf et al. [14] detected α-Syn species in live-human sPD and fPD patient plasma and cerebrospinal fluid.

2. α-Syn structure and function

Syn are a vertebrate-specific family of abundant neuronal proteins. They consist of three closely related members, α-,β-,and γ-Syn, of which α-Syn has been the prime focus ever since mutations in it were recognized as a basis for fPD. Syn is a highly conserved protein with a molecular weight of approximately 14 kDa, comprising 140 amino acids [15]. This heat-resistant, soluble, acidic protein is abundant in the presynaptic terminals of central nervous system (CNS) neurons expressed pre-dominantly in the neocortex, hippocampus, SNpc, thalamus, and cerebellum [16–18]. Unlike α- and β-synuclein, γ-Syn is not concentrated in presynaptic terminals [19] and is largely found in the peripheral nervous system (PNS).

α-Syn is composed of three distinct domains: an N-terminal amphipathic repeat region that can form α-helices; a hydrophobic central segment; and a C-terminal acidic region (**Figure 1**). The highly conserved N-terminal domain (residues 1–65) includes 6 copies of an unusual 11 amino acid repeat that display variations of a KTKEGV consensus sequence. It is unordered in solution, but can shift to a α-helical conformation [20] comprising two distinct α-helices

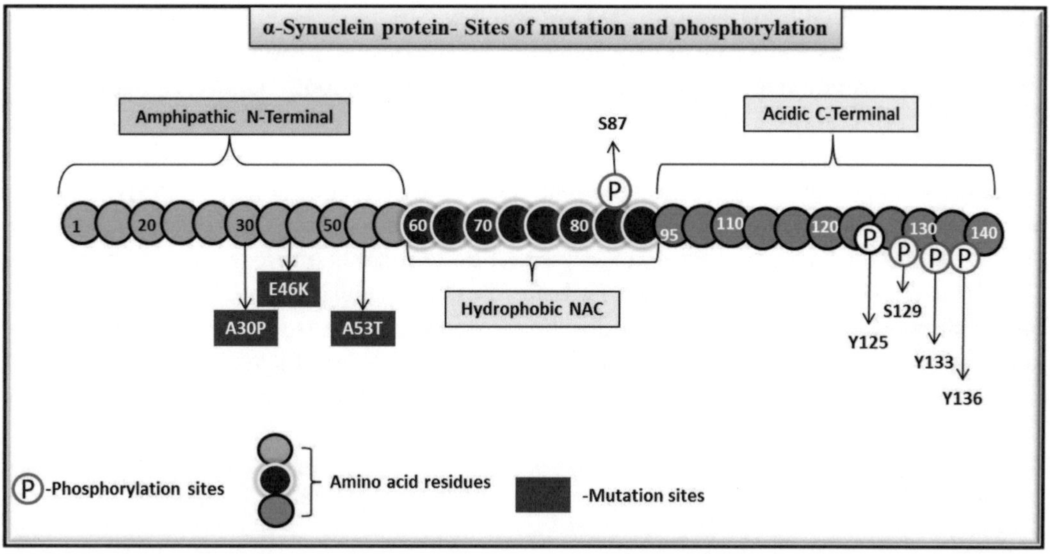

Figure 1. Schematic representation of α-Syn amino acid amino acid sequence (1–140) with amphipathic NH_2-terminal, non-amyloid component (NAC) region and acidic tail –COOH terminal. Arrows in the N-terminal, points to the three pathogenic mutations, and P in a red circle of NAC and C-terminal, points to the sites of phosphorylations.

interrupted by a short break [21]. α-Syn binds strongly to negatively charged phospholipids and becomes α-helical [22, 23], suggesting that the protein may normally be associated with the membrane [24]. This N-terminal domain includes the three sites of the fPD mutations A30P, E46K, and A53T (**Figure 1**).

The hydrophobic central segment of α-Syn (non-amyloid component (NAC), residues 66–95) (**Figure 1**) [18] is the second major component of brain amyloid plaques in Alzheimer's disease (AD) [18, 24]. This region consists of three repeats including the highly amyloidogenic part of the molecule that is responsible for α-Syn's ability to undergo a conformational change from random coil to β-sheet structure [25] and to form Aβ-like protofibrils and fibrils [24, 25]. These properties differentiate α-Syn from β-Syn and γ-Syn, which fail to form co-polymers with α-Syn [24]. The NAC region carries a phosphorylation site on Ser87 [26].

The acidic C-terminal domain (residues 96–140) of α-Syn has a strong negative charge composed primarily of acidic amino acids [20], but has no known structural elements. It consists of an acidic domain rich in proline residues (residues 125–140) that seems critical for the chaperone-like activity of α-Syn [27], as demonstrated by deletion mutants of the C-terminal region in which the α-Syn chaperone activity is lost [27–29]. In contrast to the amphipathic N-terminal and hydrophobic NAC regions, which are highly conserved between species, the C-terminal region is variable in size and in sequence [28–31]. This region is also organized in random structure in most conditions and contains several phosphorylation sites: Ser129, Tyr125, Tyr133, and Tyr136 [32] (**Figure 1**).

Although the normal functions of α-Syn are still being defined, several studies have shown that this protein has a key role to play in membrane-associated processes at the presynaptic level such as formation and maintenance of synaptic vesicle pools (**Figure 2**), regulation of lipid metabolism, and Ca^{2+} homeostasis [31–33]. Greten-Harrison et al. [34] using αβγ-Syn knockout mice has reported that deletion of Syns causes alterations in synaptic structure and transmission, age-dependent neuronal dysfunction, as well as diminished survival. *In vivo* and *in vitro* studies showed that abrogation of Syn expression decreased excitatory synapse size by ~30%, revealing that Syns are important determinants of presynaptic terminal size [34]. Younger synuclein-null mice show better basic transmission in comparison to older mice that showed a pronounced deterioration. Interestingly, it is further reported that the late onset phenotypes in Syn-null mice were not due to a loss of synapses or neurons but rather a reflection of specific changes in synaptic protein composition and axonal structure.

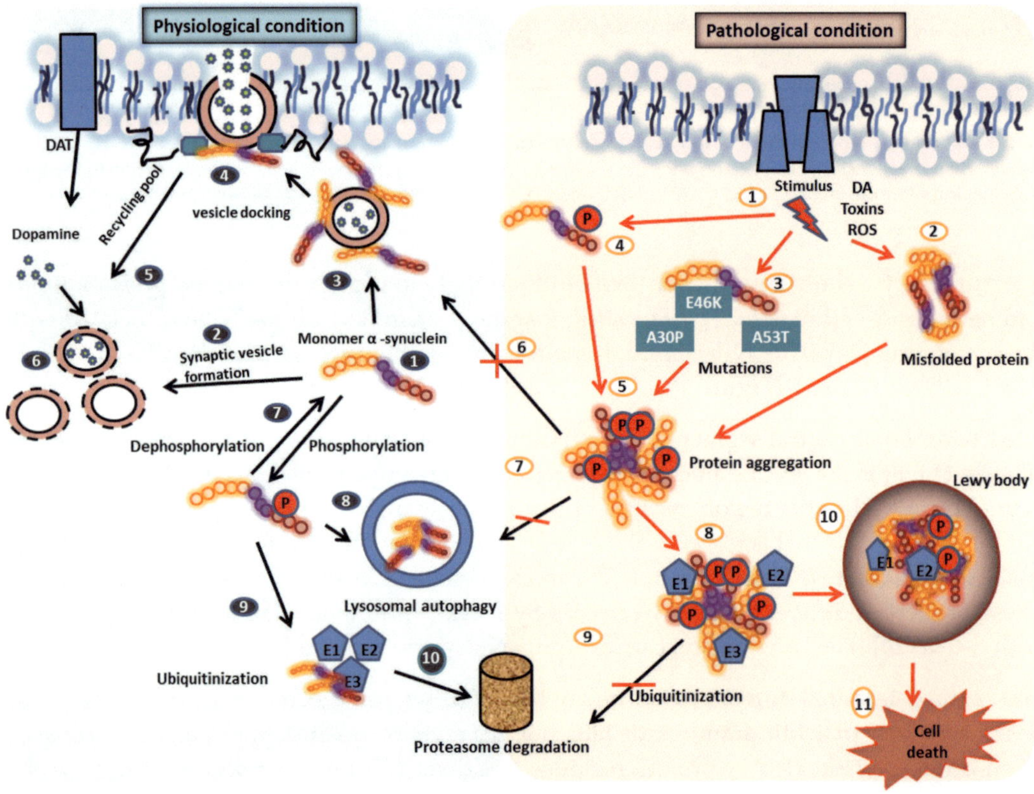

Figure 2. Schematic representation of roles of α-Syn under physiological and pathological conditions. In physiological condition, (1) α-Syn maintains synaptic functions by associating with (2) vesicle formation, (3) trafficking and (4) docking. It also associates with (5) recycling of synaptic vesicle and (6) dopamine storage. The post-translational modification of α-Syn such as (7) phosphorylation and dephosphorylation, leads to activation and deactivation of the protein. This protein undergoes (8) lysosome autophagy (9) and (10) proteosomal degradation directed by ubiquitination. However, in pathological conditions (1) due to uncertain stimulus, α-Syn undergoes (2) misfolding, (3) mutation or (4) phosphorylation leading to the (5) aggregation of the protein affecting the (6) vesicle formation, trafficking and docking, (7) impaired lysosomal autophagy, (8) and (9) ubiquitination and inhibition of proteosomal degradation. This in turn results in (10) LB formation and (11) apoptosis.

Chandra et al. [21] found selective decreases in two small synaptic signaling proteins, complexins, and 14-3-3 proteins, in α,β double-KO mice. In 2000, Abeliovich et al. [2] have shown that mice lacking α-Syn display functional deficits in the nigrostriatal dopamine system. The α-Syn−/− mice were reported to be viable and fertile, exhibited intact brain (Aβ) architecture, and possessed normal complement of DA cell bodies, fibers, and synapses; however they displayed a reduction in striatal DA and an attenuation of DA-dependent locomotor response to amphetamine [2]. Further Drolet et al.'s [35] work showed that mice lacking α-Syn have an attenuated loss of striatal dopamine following prolonged chronic 1-methyl-4-phenyl-1,2,3,6-tetrahydropyridine (MPTP) administration. In addition, ultrastructural analysis and imaging studies have shown reduced synaptic vesicle density at the active zone, and imaging further reveals a defect in the re-clustering of synaptic vesicles after endocytosis [36]. Increased levels of α-Syn thus produce specific, physiological defects in synaptic vesicle recycling that precedes detectable neuropathology.

α-Syn also has been suggested to function as a chaperone protein *in vivo*, because it appears capable of interacting with a variety of ligands and cellular proteins apart from lipids [35, 37, 38], thus modifying their activities. The N-terminal portion of α-Syn shares 40% amino acid homology with molecular chaperone 14-3-3 [36], suggesting that the two proteins could sub-serve the same function. Eliezer et al. [23] had showed that the removal of the C-terminal acidic tail of α-Syn abolished its chaperone activity. In contrast, some reports indicate that the C-terminal acidic tail is indeed necessary but not sufficient for the chaperone function of α-Syn [28, 29]. In normal physiological conditions, α-Syn exists in monomeric form and is recognized and cleared via the ubiquitin-proteasome system (UPS) and chaperone-mediated autophagy (CMA) pathways [39] (**Figure 2**). In the pathological state, misfolding or mutations of α-Syn (A30P/A53T) lead to the formation of pathologically modified species that bind with several cytoplasmic proteins and ultimately aggregate into LBs in the DA neuronal cells [40] (**Figure 2**). This aberrant level of α-Syn is also cited in idiopathic PD subjects [41].

2.1. Phosphorylation of α-synuclein at serine129

Protein phosphorylation is a reversible post-translational modification of proteins that has an important role in regulating structural and functional properties of proteins in health and disease. It is primarily associated with signaling pathways and cellular processes in all aspects of cell biology such as cell-cycle progression, differentiation, apoptosis, metabolism, transcription, cytoskeletal arrangement, intercellular communication, motility and migration [42–44]. In eukaryotes, the amino acids that are most commonly reported to be phosphorylated are serine, threonine and tyrosine [45, 46] with few reports suggesting phosphorylation at arginine, lysine and cysteine residues [45, 47, 48].

In PD too, phosphorylation appears to play an important role in fibrillogenesis, LB formation, and neurotoxicity of α-Syn *in vivo* [26, 49–51]. The majority of α-Syn in inclusions and LBs isolated from patients with PD and other synucleinopathies is phosphorylated at Ser129 (S129-P) [26, 49, 51]. Overexpression of wild-type or mutant α-Syn in *in vivo* and *in vitro* models showed expression of immunopositive phospho α-Syn serine129 (pSyn) in their proteinaceous inclusions [52–54]. Mass spectrometric analysis of α-Syn isolated from patients with synucleinopathy lesions has also confirmed that the protein is phosphorylated on Ser129 [49, 50]. More than 90% of α-Syn is phosphorylated in PD patients' brain as opposed to only 4% of phosphorylated α-Syn detected in brains of healthy subjects. Moreover, the phosphorylated α-Syn in LBs is usually ubiquitinated [50, 51, 55]. The fact that most of the α-Syn is not phosphorylated under physiological conditions *in vivo* suggests that α-Syn phosphorylation at serine129 contributes to the pathology of the disease [49, 56, 57].

2.2. Phosphorylation and CNS

The central processing unit of the human body is its CNS consisting of specialized cells called neurons relaying electrical and chemical signals to all parts of our body [58]. However, the most abundant cell type in the CNS is the glial cells comprising of astrocytes, oligodendrocytes and microglia [59, 60]. In addition, it is interspersed with microvasculature that provides the nutrients and support to these CNS cells. The central theme in CNS function is equilibrium among these various cell types to maintain optimal synaptic strengths, neuronal firing rates, and neurotransmitter release. The regulation of these functions can be either through inside-out or outside-in stimuli and are strongly associated with several signaling pathways within these cells. Virtually every class of neuronal protein is regulated by phosphorylation and most types of extracellular signals, including neurotransmitters, hormones, light, electrical potential, extracellular matrix, neurotrophic factors, and cytokines, can produce diverse physiological effects by regulating the phosphorylation of specific phosphor-proteins in their target cells. These extracellular signals modify the activity of protein kinases and/or phosphatases either directly (e.g. receptors with kinase activity) or via cascades of enzymatic reactions (e.g. receptor → G protein → enzyme ~ second messenger ~ protein kinase).

2.3. Kinases involved in α-synuclein phosphorylation

α-Syn phosphorylation can be induced by several kinases. Serine129 of α-Syn can be phosphorylated by G protein-coupled receptor kinases (GRK1, GRK2, GRK5, and GRK6) [61–63], casein kinases 1 and 2 (CK1 and CK2) [26, 64–69], and the polo-like kinases (PLKs) [70]. Current studies have shown that, GRKs may also phosphorylate non-receptor substrates, comprising the four members of the Syn family (α-,β-,γ-Syn, and synoretin) in addition to phosphorylating agonist-occupied G protein-coupled receptors (GPCRs) [62]. Overexpression of GRK2 or GRK5 in COS-1 cells, showed that these kinases phosphorylate α-Syn at serine129 [62]. Endogenous GRK-induced phosphorylation of α-Syn at serine129 was demonstrated *in vitro* in HEK293 cells, and GRK3 and GRK6 were seen to be playing the main roles in this modification [63]. Post-mortem analysis showed that GRK5 co-localized with α-Syn in the LBs of the SNpc of PD patients, but was not detected in cortical LBs of Dementia with lewy bodies (DLB), or in the glial cytoplasmic inclusions of MSA [61]. Overexpression of α-Syn in SH-SY5Y cells and human α-Syn expressing transgenic mice also showed an increase in GRK5 protein [71]. A genetic association study revealed a haplotype association of the GRK5 gene with susceptibility to sPD in the Japanese population [61]; however, GRK5 polymorphisms in southern Italy failed to correlate with sPD [72]. The knockdown of endogenous GRK5 in SH-SY5Y cells fails to suppress the phosphorylation of α-Syn completely [71], confirming the involvement of other kinases in this phosphorylation.

The other group of kinases that phosphorylates α-Syn at serine129 is CK1 and CK2. This has been demonstrated in the yeast model [69], in mammalian cells [26, 68] and in rat primary cortical neurons [64]. It has been suggested that phosphorylation of serine129 in α-Syn by CK2 may promote *in vitro* fibrillation [59] and *in situ* inclusion formation [73]. Phosphorylation by CK2 and dephosphorylation by PP2C in *in vitro* model indicate that these may be important enzymes that regulate the phosphorylation of α-Syn [74]. Furthermore, surplus α-Syn can form inclusions that sequester CK1, diminishing CK1 activity and aggravating synaptic defects, generating a vicious toxic cycle. CK1 has been found to co-localize with pS87 in transgenic mice and in LB-like structures in LBD/PD-diseased brains [75, 76], and phosphorylates α-Syn at serine87 as well [26]. Oxidative stress induced by iron is reported to upregulate CK2 that leads to increased phosphorylated α-Syn serine129 with an associated increase in oligomerization and inclusion formation [65]. Smith et al. [66] have shown in SH-SY5Ycells that the increase in α-Syn phosphorylation under oxidative stress is mediated by CK2 and correlates with enhancement of inclusion formation.

In vitro studies employing kinase assays showed involvement of another family of cellular kinases in the phosphorylation of α-Syn at serine129: PLK1, PLK2, and PLK3 [70, 77]. The PLKs comprise a family of conserved Ser/Thr protein kinases that are known to play a role in cell-cycle regulation and cellular response to stress and carcinogenesis [78]. PLK2 directly phosphorylated α-Syn at serine129 in an *in vitro* biochemical assay [70]. Inhibitors of PLK kinases inhibited α-Syn phosphorylation both in primary cortical cell cultures and in mouse brain *in vivo*. Further, using PLK2 KO mouse, Buck et al. [79] too have shown that PLK2 plays a key role in Ser129 α-Syn phosphorylation in mouse brain. Aubele et al. [80] have shown in an *in vivo* model that brain-permeable Plk-2 inhibitors reduce α-Syn phosphorylation in rat brain. In response to synaptic activation, PLK2 and PLK3 expression is reported that is associated with synaptic plasticity, remodeling, and homeostasis [81, 82], thus suggesting that these kinases could play an important role in modulating the normal physiology of α-Syn. Leucine-rich repeat kinase 2 (LRRK2) is also known to pSyn [83], but this remains debatable as there are no other studies confirming this, despite the existence of a clear interaction between the two proteins [83, 84].

2.4. Phosphorylation at serine129 modulates α-synuclein protein-protein interaction

The C-terminal domain of α-synuclein (residues 96–140) is an acidic tail of 43AA residues, containing 10 Glu and 5 Asp residues. C-terminal truncations of α-Syn induce aggregation, suggesting that C-terminal modifications might be involved in the pathology of α-Syn [85]. An interaction between the C-terminal domain and the NAC region of α-Syn is postulated to be responsible for the inhibition of α-Syn aggregation. Moreover, there are several studies on the interaction of α-Syn C-terminal tail with different proteins [86–91]. McFarland et al. [92] were the pioneers to address this using targeted functional proteomics [51]. The authors showed that the non-phosphorylated α-Syn peptide primarily interacts with proteins related to the mitochondrial electron transport chain (ETC) (complex I, III, and IV proteins of the ETC) [51]. It was hypothesized that changes in α-Syn phosphorylation could represent a response to biochemical events associated with PD pathogenesis. Among these, mitochondrial complex I dysfunction, oxidative stress, and proteasome dysfunction are processes that are known to be involved in synucleinopathies [93, 94]. The low levels of pSyn under physiological conditions as well as the absence of other phosphorylated residues such as pY39, pS87 and pY125 [26, 49, 50] suggest a faster degradation of this form under normal conditions. In fact, the phosphorylation status of α-Syn was recently correlated with clearance mechanisms [95, 96]. Another group Chau et al. [97] too reported that α-Syn phosphorylation at serine129 is toxic to DA cells and both the levels of serine129 phosphorylated protein as well as its toxicity are increased with proteosomal inhibition, emphasizing the interdependence of these pathways in PD pathogenesis.

However, the pSyn has a greater affinity for certain cytoskeletal and presynaptic proteins associated with synaptic transmission and vesicle trafficking [51]. Yin et al. [98] showed that α-Syn interacts with the switch region of Rab8a, a small guanine nucleotide-binding protein, in a serine129 phosphorylation-dependent manner; thus implicating its role in coordinating vesicle trafficking. Hara et al. [99] reported that serine129 phosphorylation of membrane-associated α-Syn modulates dopamine transporter function in a G protein-coupled receptor kinase-dependent manner. These observations suggest that pSyn could serve as a molecular switch to control α-Syn interaction with different protein partners and therefore may modulate the function of DA neurons. However, further investigations are required to assess the impact and the physiological consequences of serine129 phosphorylation on α-Syn interaction with other proteins, such as SNARE proteins [35, 100], cytoskeletal proteins (i.e. tubulin) [55, 101] and other amyloidogenic proteins (i.e. tau) [19, 102]. Jensen et al. [103] have hypothesized that an interaction between α-Syn and tau could link synaptic vesicles with microtubules. Tau has been shown to co-localize and interact directly with the Src PTK family member, Fyn [104]. It is hypothesized that tau could bring Src PTK family members such as Fyn into closer proximity to α-Syn, thereby enhancing the activity of these kinases for α-Syn. Samuel et al. [105] showed that the membrane binding of α-Syn monomers was differentially affected by phosphorylation depending on the PD-linked mutation. WT α-Syn binding to presynaptic membranes was not affected by phosphorylation, while A30P α-Syn binding was greatly increased and A53T α-Syn was marginally lower, implicating the distal effects of the carboxyl terminal on amino-terminal membrane binding. The un-phosphorylated form of serine129 associates mainly with mitochondrial electron transport proteins, while the phosphorylated form associates with cytoskeletal, vesicular trafficking proteins and enzymes involved in protein serine phosphorylation [92]. Further work by Sugeno et al. [106] using α-Syn-over expressing cells exposed to a low dose of rotenone as an environmental toxin, showed that phosphorylation of α-Syn at serine129 promoted intracellular aggregate-formation and induced ER stress that was followed by mitochondrial damage and apoptosis.

Phosphorylation also seems to play an important role in the regulation of α-Syn axonal transport as the serine129D mutation significantly decreases its rate of transport in neurons, probably due to the modulation of α-Syn interaction with motor and/or accessory proteins involved in this process [107]. Moreover, the interplay between the different phosphorylated residues could increase the diversity in the possible protein interactors. Several differences were observed in the set of proteins that were found to interact with serine129 and Y125-phosphorylated forms of α-Syn [92]. S129 and Y125 residues both residing in the C-terminal region of α-Syn have been implicated in the majority of α-Syn interactions with proteins

[103, 108, 109], reinforcing the significance of phosphorylation in these residues in modulating the biological role of α-Syn. All these findings together suggest that phosphorylation of α-Syn at serine129 has a widespread effect on protein-protein interaction of α-Syn.

Phosphorylation also seems to alter the subcellular localization of α-Syn. pSyn was found to be preferentially localized in the nuclei of DA neurons in rat and mouse models of synucleinopathy [67, 100]. In studies using PD rat models, the phospho-resistant S129A was found to be localized in the nucleus at higher levels than the S129D form, and was found to correlate with enhanced toxicity [110, 111]. Our group too demonstrated the nuclear localization of pSyn in SH-SY5Y cells under 6-hydroxydopamine toxicity [112]. Gonçalves and Outeiro [113] showed that S129 phosphorylation modulates the shuttling of α-Syn between nucleus and cytoplasm in human neuroglioma cells, using photo-activatable green fluorescent protein as a reporter. Moreover, the study showed that co-expression of α-Syn with different kinases altered the translocation dynamics of the protein. While G protein-coupled receptor kinase 5 (GRK5) promotes the nuclear localization of α-Syn, PLK2 and three modulate the shuttling of the protein between the nucleus and cytoplasm [114]. This difference reflects different α-Syn phosphorylation patterns in serine129 and/or other residues. Although the function of α-Syn in the nucleus is still unclear, it appears to be related to pathological insults. In particular, nuclear localization of α-Syn increases under oxidative stress conditions [112, 114, 115]. Nuclear α-Syn interacts with histones, inhibits acetylation, and promotes neurotoxicity [116, 117]. Furthermore, α-Syn may act as a transcriptional regulator, binding promoters such as PGC1-α, a master regulator of mitochondrial gene expression [106]. The significance of pSyn in regulating nuclear proteins still needs to be unraveled.

2.5. Oxidative stress and α-Syn phosphorylation

Oxidative stress can increase α-Syn phosphorylation [97]. Perfeito et al. [118] showed through an *in vitro* model that exposure to ferrous iron and rotenone resulted in increase in pSyn. A similar increase was reported by Ganapathy et al. [112] in the presence of the endogenous toxin 6-hydroxydopamine. Proteosomal inhibition by epoxomycin and increased oxidative stress by paraquat treatment has led to increases in pSyn [97]. This may reflect either an increased activity of the kinase responsible or a decrease of phosphatase activity. Under physiological conditions, α-Syn is degraded by chaperone-mediated autophagy [119], and studies suggest that this gets reduced upon oxidation or nitration [12, 106]. α-Syn may also be degraded by proteasomes [120]. The increase in pSyn levels in LBs suggests a change in the turnover or degradation of the phosphorylated protein (**Figure 3**). It is possible that at elevated levels, this phosphorylated species may be toxic.

2.6. α-Syn phosphorylation at serine129 and cellular events

The role of α-Syn phosphorylation in the cellular pathogenesis of PD remains debated [92, 110, 111, 121, 122]. This apparent controversy is due to the fact that phosphomimics (S129D/E) do not reproduce the exact properties of the endogenous authentically phosphorylated α-Syn [122–124]. The expression of a variant showing prevention of phosphorylation by site-directed mutagenesis of serine129 to alanine (S129A) caused an increase in α-Syn inclusions

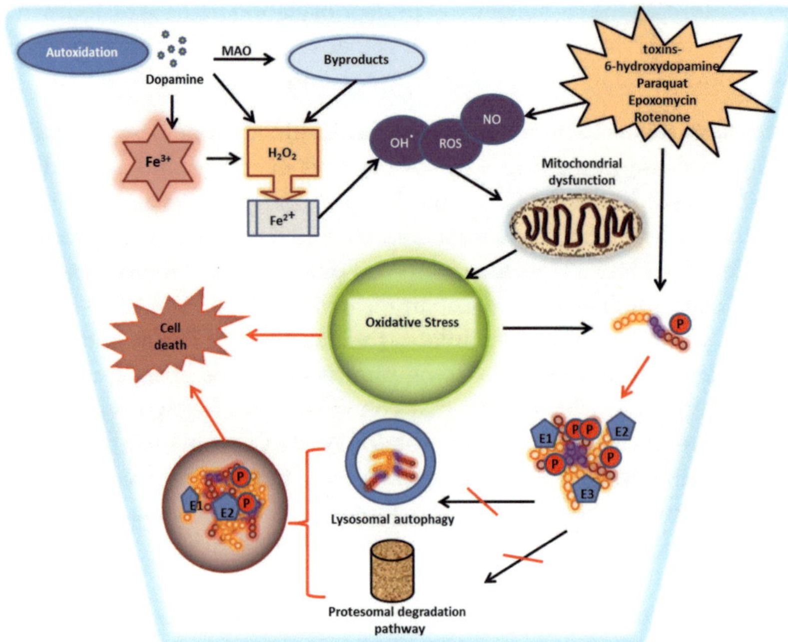

Figure 3. Schematic representation of suggested pathological role of oxidative stress in the α-Syn phosphorylation and aggregation. Normally genetic mutation, neurotoxins, dopamine auto oxidation in the presence of excess iron and the release of reactive oxygen species (ROS) leads to mitochondrial dysfunction and oxidative stress in turn leading to cell death. However, increase in the oxidative stress and direct effect of neurotoxins can lead to phosphorylation of α-Syn and their aggregation which inturn inhibits the lysosomal autophagy and impaired proteosomal degradation resulting in LB formation and cell death.

and toxicity [110] On the other hand, several investigators have reported that the expression of S129A mutant protein led to fewer inclusions [66, 94, 110]. It has also been reported that the expression of phosphorylation mimicking serine129 to the aspartate (S129D) variant does not show DA deficits [92, 110, 111, 121, 123]. Hence, several groups have sought to address the issue by overexpressing α-Syn and using siRNA for its natural kinases [99, 118, 124]. As discussed in the earlier section, among the kinases primarily responsible for the α-Syn phosphorylation at serine129 are CKs, GRKs, LRRK2, and PLKs. The modulation of phosphorylation of α-Syn has also been reported by targeting the kinases in A53T mutant α-Syn expressing cells [99, 118, 124–126]. siRNA studies targeting kinases such as PLK, GRK2, and CK2 have been used to study the effect of phosphorylated α-synuclein on WT and A53T mutant α-synuclein expressing cells [99, 118, 124, 125, 127]. These studies have shown the effect of α-Syn phosphorylation with respect to ROS generation, mitochondrial alterations, proteasomal changes, and dopamine transport. The work of Perfeito et al. [118] suggest that stimuli that promote ROS formation and mitochondrial alterations highly correlate with mutant α-Syn phosphorylation at serine129, which may precede cell degeneration in PD. Similarly, we have shown in our recent work that at sub-lethal 6-hydroxydopamine concentrations, the decrease in resting vesicles (VMAT2) and vesicular dopamine release are not attributable to apoptotic cell death and occur concomitantly with the phosphorylation of α-Syn [112].

2.7. α-Syn phosphorylation at serine129 during PD pathogenesis: an early or late event?

The evidence of pSyn accumulation in the brain has been collected largely from post mortem analysis and it fails to answer if this accumulation occurs during the early or late stages of synucleinopathies. In a recent work, Walker et al. [127] investigated how pSyn levels and solubility change in cingulate and temporal cortex of DLB patients, at different stages of the disease. The authors reported a progressive accumulation of pSyn-immunoreactive species in diseased brains compared to healthy controls, as well as a positive correlation between pSyn levels and the severity of disease symptoms. A similar study using brain samples from PD patients also reported a drastic accumulation of pSyn-positive inclusions in different brain regions at the late stages of the disease [128]. Together, these results suggest that the occurrence of pSyn is linked to the severity of disease progression.

In our recent work, we demonstrated using biophysical and biochemical analysis in an *in vitro* model that under sub-lethal concentrations of 6-hydroxydopamine, phosphorylation of α-Syn precedes apoptosis and occurs concomitantly with the decrease in vesicular dopamine release [21]. This study provides a new perspective on the occurrence of pSyn even in early stages of the disease that may contribute to the impairment of neuronal function. Another recent work by Takahashi et al. [53] demonstrated increased accumulation of pSyn in the motor cortex of normal aging mice along with early onset motor impairment, which was further ameliorated by coenzyme Q.

2.8. pSyn in human fluids and PNS of synucleinopathies

Several studies in the field of PD and diagnosis of Parkinsonism are based on the fundamental molecular events associated with or without LBs. Total α-Syn quantification in the CSF of PD, DLB, and MSA patients in comparison to healthy controls has been proposed as a biomarker for α-Syn-related disorders [129–131]. However, an ideal biomarker for a particular disease must be easy to detect, and also reflect disease onset and progression with associated primary changes. A longitudinal study conducted by Foulds et al. [132] in the blood plasma of patients suffering from PD showed that total α-Syn in blood plasma of PD patients remained similar to that in normal individuals, but the level of α-Syn phosphorylated at serine129 was significantly higher in PD patients. Statistical analysis confirmed the usefulness of plasma levels of pSyn in discriminating patients with PD from healthy controls. In addition, pSyn inclusions were also detected in the PNS which might also serve as a useful diagnostic test for PD and related synucleinopathies. Work from two independent groups on skin biopsies showed that the majority of PD patients had accumulation of pSyn in small and large nerve fibers, while no signal was detected in healthy controls and in MSA or essential tremor control subjects [133, 134]. This cutaneous pathology was correlated with the progression of disease symptoms suggesting the use of this peripheral marker as a biomarker for the disease [135]. Presence of pSyn immunoreactivity was detected in gastric, duodenal and colonic biopsies [136, 137]. Hilton et al. [137] further reports that this accumulation of pSyn in the bowels of patients is detected in the pre-clinical phase of PD. Taken together, these reports suggest that accumulation of pSyn might provide a more reliable biomarker to detect PD at early stages and further discriminate between synucleinopathies, compared to total α-Syn.

3. Conclusion

In comparison to the relatively small number of genes, the large diversity of the proteome is achieved mainly by post-translational protein modifications, phosphorylation being the most widespread. Since the C-terminal region of α-Syn is involved in interaction with proteins [103, 104, 109] and metal ions [138–140], any phosphorylation in this site can alter its interaction capability [77]. The significance of α-Syn phosphorylation at serine129 has gained importance in PD because its accumulation is distinctly enhanced in the diseased condition. The revelation of the involvement of pSyn on α-Syn aggregation, LB formation, and neurotoxicity is crucial to understanding the pathogenesis and progression of PD and related disorders. Since some *in vitro* and *in vivo* studies have indicated that the pSyn is an early event preceding apoptosis, some important questions now needs to be explored in reference to the physiological functions regulated by phosphorylation such as dopamine synthesis, vesicle mobilization, and regulation of synaptic proteins, synaptic plasticity, and its subcellular localization. A systemic investigation of the role of each kinase on the phosphorylation of α-Syn and α-Syn clearance also needs to be carried out to identify viable targets for development of new therapeutics. A further *in vivo* study by Hirai et al. [57] elevates the significance of pSyn as a region-specific phenomenon, and the level of pSyn in response to physiological stimuli is selectively higher in the striatum region in comparison to cortex and hippocampus [57]. The relative sensitivity of this phenomenon in the striatum in reference to PD pathology needs to be assessed to address the increased susceptibility of some brain regions to α-Syn phosphorylation and pathology. In addition, the synergistic association of PD-linked mutations and pSyn in the regulation of α-Syn toxicity needs to be evaluated. Investigation of the possible implications of phosphorylation of α-Syn on cell-to-cell transmission and its pathological propagation in PD-diseased brains is also pending. Thus, the research evidence clearly suggests that the phosphorylation of α-Syn plays a significant and possibly pivotal role both in PD pathogenesis and progression, and that answers to the questions above are crucial for the identification of novel, disease-modifying, therapeutic targets for the treatment of PD and related synucleinopathies.

Author details

Indrani Datta* and Kavina Ganapathy

*Address all correspondence to: indranidatta.nimhans@gmail.com

Department of Biophysics, National Institute of Mental Health and Neurosciences, Bengaluru, India

References

[1] De Lau LM, Breteler MM. Epidemiology of Parkinson's disease. The Lancet Neurology. 2006;**5**(6):525-535

[2] Abeliovich A, Schmitz Y, Farinas I, Choi-Lundberg D, Ho WH, et al. Mice lacking α-synuclein display functional deficits in the nigrostriatal dopamine system. Neuron. 2000;**25**:239-252

[3] Alexander GE. Biology of PD: Pathogenesis & pathophysiology of a multisystem neurodegenerative disorder. Dialogues in Clinical Neuroscience. 2004;**6**(3):259-280

[4] Gesi M, Soldani P, Giorgi FS, Santinami A, Bonaccorsi I, Fornai F. The role of the locus coeruleus in the development of PD. Neuroscience & Biobehavioral Reviews. 2000;**24**(6):655-668

[5] Hornykiewicz O. Parkinson's disease and the adaptive capacity of the nigrostriatal dopamine system: Possible neurochemical mechanisms. Advances in Neurology1993; **60**:140-147

[6] Rodriguez-Oroz MC, Jahanshahi M, Krack P, Litvan I, Macias R, et al. Initial clinical manifestations of PD: Features and pathophysiological mechanisms. The Lancet Neurology. 2009;**8**(12):1128-1139

[7] Polymeropoulos MH, Lavedan C, Leroy E, Ide SE, Dehejia A, et al. Mutation in the α-synclein gene identified in families with PD. Science. 1997;**276**:2045-2047

[8] Kruger R, Kuhn W, Mller T, Woitalla D, Graeber M, Kösel S, et al. Ala53Pro mutation in the gene encoding α-synuclein in PD. Nature Genetics. 1998;**18**:106-108

[9] Zarranz JJ, Alegre J, Gómez-Esteban JC, Lezcano E, Ros R, et al. The mutation, E46K of α-synuclein causes Parkinson and Lewy body dementia. Annals of Neurology. 2004;**55**:164-173

[10] Singleton AB, Farrer M, Johnson J, Singleton A, Haugue S, et al. α-Synuclein locus triplication causes PD. Science. 2003;**302**:841

[11] Masliah E, Rockenstein E, Veinbergs I, Mallory M, Hashimoto M, et al. Dopaminergic loss and inclusion body formation in α-synuclein mice. Implications for neurodegenerative disorders. Science. 2000;**287**:1265-1269

[12] Scherzer CR, Grass JA, Liao Z, Pepivani I, Zheng B, et al. GATA transcription factors directly regulate the PD-linked gene α-synuclein. Proceedings of the National Academy of Sciences of the United States of America. 2008;**105**:10907-10912

[13] Martinez-Vicente M, Talloczy Z, Kaushik S, Massey AC, Mazzulli J, et al. Dopamine-modified α-synuclein blocks chaperone-mediated autophagy. The Journal of Clinical Investigation. 2008;**118**(2):777-788

[14] El-Agnaf OMA, Salem SA, Paleologou KE, Curran MD, Gibson MJ, et al. Detection of oligomeric forms of α-synuclein protein in human plasma as a potential biomarker for PD. The FASEB Journal. 2006;**20**:419-425

[15] Beyer K. α-Synuclein structure, posttranslational modification and alternative splicing as aggregation enhancers. Actaneuropathologica. 2006;**112**(3):237-251

[16] Iwai A, Masliah E, Yoshimoto M, Ge N, Flanagan L, de Silva HA, Kittel A, Saitoh T. The precursor protein of non-A component of Alzheimer's disease amyloid is a presynaptic protein of the central nervous system. Neuron. 1995;**14**:467-475

[17] Waxman EA, Giasson BI. Molecular mechanisms of α-synuclein neurodegeneration. Biochimicaet Biophysica Acta (BBA) – Molecular Basis of Disease. 2009;**1792**(7):616-624

[18] Uéda K, Fukushima H, Masliah E, Xia Y, Iwai A, et al. Molecular cloning of cDNA encoding an unrecognized component of amyloid in Alzheimer disease. Proceedings of the National Academy of Sciences of the United States of America. 1993;**90**:11282-11286

[19] Jakes R, Spillantini MG, Goedert MI. Identification of two distinct synucleins from human brain. FEBS Letters. 1994;**345**:27-32

[20] George JM. The synucleins. Genome Biology. 2002;**3**:3002-3006

[21] Chandra S, Chen X, Rizo J, Jahn R, Südho TC. A broken-helix in folded α-synuclein. Journal of Biological Chemistry. 2003;**278**:15313-15318

[22] Davidson WS, Jonas A, Clayton DF, George JM. Stabilization of α-synuclein secondary structure upon binding to synthetic membranes. Journal of Biological Chemistry. 1998;**273**:9443-9449

[23] Eliezer D, Kutluay E, Bussell R, Jr, Browne G. Conformational properties of α-synuclein in its free and lipid-associated states. Journal of Molecular Biology. 2001;**307**:1061-1073

[24] Giasson BI, Murray IVJ, Trojanowski JQ, Lee VM. A hydrophobic stretch of 12 amino acid residues in the middle of synuclein is essential for filament assembly. Journal of Biological Chemistry. 2001;**276**:2380-2386

[25] El-Agnaf OMA, Irvine GB. Aggregation and neurotoxicity of α-synuclein and related peptides. Biochemical Society Transactions.2002;**30**:559-565

[26] Okochi M, Walter J, Koyama A, Nakajo S, Baba M, et al. Constitutive phosphorylation of the PD associated α-synuclein. Journal of Biological Chemistry. 2000;**275**:390-397

[27] Souza JM, Giasson BI, Lee VM, Ischiropoulos H. Chaperone-like activity of synucleins. FEBS Letters. 2000;**474**:116-119

[28] Kim TD, Paik SR, Yang CH. Structural and functional implications of C-terminal regions of α-synuclein. Biochemistry. 2002;**41**:13782-13790

[29] Park SM, Jung HY, Kim TD, Park JH, Yang CH, Kim, J. Distinct roles of the N-terminal-binding domain and the C-terminal-solubilizing domain of α-synuclein, a molecular chaperone. Journal of Biological Chemistry. 2002;**277**:28512-28520

[30] Uversky VN. Neuropathology, biochemistry, and biophysics of synuclein aggregation. Journal of Neurochemistry. 2007;**103**:17-37

[31] Lavedan C. The synuclein family. Genome Research. 1998;**8**:871-880

[32] Cavallarin N, Vicario M, Negro A. The role of phosphorylation in synucleinopathies: Focus on PD. CNS & Neurological Disorders-Drug Targets. 2010;**9**(4):471-481.

[33] Auluck PK, Caraveo G, Lindquist S. α-Synuclein: Membrane interactions and toxicity in PD. Annual Review of Cell and Developmental Biology. 2010;**26**:211-233

[34] Greten-Harrison B, Polydoro M, Morimoto-Tomita M, Diao L, Williams AM, et al. αβγ-Synuclein triple knockout mice reveal age-dependent neuronal dysfunction. Proceedings of the National Academy of Sciences. 2010;**107**(45):19573-19578

[35] Drolet RE, Behrouz B, Lookingland KJ, Goudreau JL. Mice lacking α-synuclein have an attenuated loss of striatal dopamine following prolonged chronic MPTP administration. Neurotoxicology. 2004;**25**(5):761-769

[36] Nemani VM, Lu W, Berge V, Nakamura K, Onoa B, et al. Increased expression of α-synuclein reduces neurotransmitter release by inhibiting synaptic vesicle reclustering after endocytosis. Neuron. 2010;**65**(1):66-79

[37] Xu J, Kao S-Y, Lee FJ, Song W, Jin L-W, Yankner BA. Dopamine-dependent neurotoxicity of α-synuclein: A mechanism for selective neurodegeneration in PD. Nature Medicine. 2002;**8**:600-606

[38] Perez RG, Waymire JC, Lin E, Liu JJ, Guo F, Zigmond MJ. A role for α-synuclein in the regulation of dopamine biosynthesis. Journal of Neuroscience. 2002;**22**:3090-3099

[39] Mazzulli JR, Zunke F, Tsunemi T, Toker NJ, Jeon S, et al. Activation of β-glucocerebrosidase reduces pathological α-synuclein and restores lysosomal function in Parkinson's patient midbrain neurons. Journal of Neuroscience.2016;**36**(29):7693-7706

[40] Janda E, Isidoro C, Carresi C, Mollace V. Defective autophagy in PD: Role of oxidative stress. Molecular Neurobiology. 2012;**46**(3):639-661

[41] McCormack A, Chegeni N, Chegini F, Colella A, Power J, Keating D, Chataway T. Purification of α-synuclein containing inclusions from human post mortem brain tissue. Journal of Neuroscience Methods. 2016;**266**:141-150

[42] Manning G, Plowman GD, Hunter T, Sudarsanam S. Evolution of protein kinase signaling from yeast to man. Trends in Biochemical Sciences. 2002;**27**:514-520

[43] Manning G, Whyte DB, Martinez R, Hunter T, Sudarsanam S. The protein kinase complement of the human genome. Science. 2002;**298**:1912-1934

[44] Navarro J, del Moral R, Marijuán PC. Charting the Signaling Pathways of the Neuron. The Wiley Handbook of Evolutionary Neuroscience; first edition, John wiley &sons Ltd. 2016 Dec 12:49.

[45] Cieśla J, Frączyk T, Rode W. Phosphorylation of basic amino acid residues in proteins: Important but easily missed. Acta Biochimica Polonica. 2011;**58**(2):137-148

[46] Deutscher J, Saier Jr MH. Ser/Thr/Tyr protein phosphorylation in bacteria–for long time neglected, now well established. Journal of Molecular Microbiology and Biotechnology. 2006;**9**(3-4):125-131

[47] Fuhrmann J, Subramanian V, Kojetin DJ, Thompson PR. Activity-based profiling reveals a regulatory link between oxidative stress and protein arginine phosphorylation. Cell Chemical Biology. 2016;**23**(8):967-977

[48] Matthews HR. Protein kinases and phosphatases that act on histidine, lysine, or arginine residues in eukaryotic proteins: A possible regulator of the mitogen-activated protein kinase cascade. Pharmacology & Therapeutics. 1995;**67**(3):323-350

[49] Anderson JP, Walker DE, Goldstein JM, de Laat R, Banducci K, et al. Phosphorylation of Ser-129 is the dominant pathological modification of α-synuclein in familial and sporadic Lewy body disease. Journal of Biological Chemistry. 2006;**281**:29739-29752

[50] Fujiwara H, Hasegawa M, Dohmae N, Kawashima A, Masliah E, Goldberg MS, et al. α-Synuclein is phosphorylated in synucleinopathy lesions.Nature Cell Biology. 2002;**4**:160-164

[51] Kahle PJ, Neumann M, Ozmen L, Muller V, Jacobsen H, et al. Subcellular localization of wild-type and Parkinson's disease-associated mutant α-synuclein in human and transgenic mouse brain. Journal of Neuroscience. 2000;**20**:6365-6373

[52] Chen L, Feany MB. α-Synuclein phosphorylation controls neurotoxicity and inclusion formation in a Drosophila model of PD. Nature Neuroscience. 2005;**8**:657-663

[53] Takahashi M, Kanuka H, Fujiwara H, Koyama A, Hasegawa M, et al. Phosphorylation of α-synuclein characteristic of synucleinopathy lesions is recapitulated in α-synuclein transgenic Drosophila. Neuroscience Letters. 2003;**336**:155-158

[54] Yamada M, Iwatsubo T, Mizuno Y, Mochizuki H. Overexpression of α-synuclein in rat SNpc results in loss of dopaminergic neurons, phosphorylation of α-synuclein and activation of caspase-9: Resemblance to pathogenetic changes in PD. Journal of Neurochemistry. 2004;**91**:451-461

[55] Hasegawa M, Fujiwara H, Nonaka T, Wakabasyashi KK, Takahashi H, et al. Phosphorylated α-synuclein is ubiquitinated in synucleopathy lesions. Journal of Biological Chemistry. 2002;**277**:49071-49076

[56] Arima K, Hirai S, Sunohara N, Aoto K, Izumiyama Y, et al. Cellular co-localization of phosphorylated tau- & NACP/-synuclein-epitopes in Lewy bodies in sporadic PD and in dementia with Lewy bodies. Brain Research. 1999;**843**:153-161

[57] Hirai Y, Fujita SC, Iwatsubo T, Hasegawa M. Phosphorylated α-Syn in normal mouse brain. FEBS Letters. 2004;**572**:227-232

[58] Sherwood L. Human Physiology: From Cells to Systems. Cengage learning Brooks/Cole; 2015. ISBN-13: 978-0-495-39184-5

[59] Dobson KL, Bellamy TC. Glial Cells. In Essentials of Cerebellum and Cerebellar Disorder. Springer International Publishing. 2016;219-223.

[60] Peters DG, Connor JR. Introduction to cells comprising the nervous system In: Glycobiology of the Nervous System. New York: Springer. 2014. pp. 33-45

[61] Arawaka S, Wada M, Goto S, Karube H, Sakamoto M, Ren CH, et al. The role of G-protein-coupled receptor kinase 5 in pathogenesis of sporadic PD. Journal of Neuroscience. 2006;**26**:9227-9238

[62] Pronin AN, Morris AJ, Surguchov A, Benovic JL. Synucleins are a novel class of substrates for G proteincoupled receptor kinases. Journal of Biological Chemistry. 2000;**275**:26515-26522

[63] Sakamoto M, Arawaka S, Hara S, Sato H, Cui C, Machiya Y, et al. Contribution of endogenous G-protein-coupled receptor kinases to Ser129 phosphorylation of α-synuclein in HEK293 cells. Biochemical and Biophysical Research Communications. 2009;**384**:378-382

[64] Ishii A, Nonaka T, Taniguchi S, Saito T, Arai T, Mann D, et al. Casein kinase 2 is the major enzyme in brain that phosphorylates Ser129 of human α-synuclein: Implication for α-synucleinopathies. FEBS Letters. 2007;**581**:4711-4717

[65] Takahashi M, Ko LW, Kulathingal J, Jiang P, Sevlever D, Yen SH. Oxidative stress-induced phosphorylation, degradation and aggregation of α-synuclein are linked to upregulated CK2 & cathepsin D. European Journal of Neuroscience. 2007;**26**:863-874

[66] Smith WW, Margolis RL, Li X, Troncoso JC, Lee MK, Dawson VL, et al. α-synuclein phosphorylation enhances eosinophilic cytoplasmic inclusion formation in SH-SY5Y cells. Journal of Neuroscience. 2005;**25**:5544-5555

[67] Wakamatsu M, Ishii A, Ukai Y, Sakagami J, Iwata S, Ono M, et al. Accumulation of phosphorylated α-synuclein in dopaminergic neurons of transgenic mice that express human α-synuclein. Journal of Neuroscience Research. 2007;**85**:1819-1825

[68] Waxman EA, Giasson BI. Specificity and regulation of casein kinase-mediated phosphorylation of α-synuclein. Journal of Neuropathology & Experimental Neurology. 2008;**67**:402-416

[69] Zabrocki P, Bastiaens I, Delay C, Bammens T, Ghillebert R, Pellens K, et al. Phosphorylation, lipid raft interaction and traffic of α-synuclein in a yeast model for Parkinson. Biochimica et Biophysica Acta. 2008;**1783**:1767-1780

[70] Inglis KJ, Chereau D, Brigham EF, Chiou SS, Schobel S, Frigon NL, et al. Polo-like kinase 2 (PLK2) phosphorylates α-synuclein at serine 129 in central nervous system. Journal of Biological Chemistry. 2009;**284**:2598-2602

[71] Liu P, Wang X, Gao N, Zhu H, Dai X, Xu Y, et al. G protein-coupled receptor kinase 5, overexpressed in the α-synuclein up-regulation model of PD, regulates bcl-2 expression. Brain Research. 2010;**1307**:134-141

[72] Tarantino P, De Marco EV, Annesi G, Rocca FE, Annesi F, Civitelli D, et al. Lack of association between G-protein coupled receptor kinase5 gene & PD. American Journal of Medical Genetics Part B: Neuropsychiatric Genetics. 2011;**156B**:104-107

[73] Nishie M, Mori F, Fujiwara H, Hasegawa M, Yoshimoto M, Iwatsubo T, et al. Accumulation of phosphorylated α-synuclein in the brain and peripheral ganglia of patients with multiple system atrophy. Acta Neuropathologica. 2004;**107**:292-298

[74] Waxman EA, Giasson BI. Characterization of kinases involved in the phosphorylation of aggregated α-synuclein. Journal of Neuroscience Research. 2011;**89**:231-247

[75] Ryu, MY, Kim DW, Arima K, Mouradian MM, Kim SU, and Lee G. Localization of CKII beta subunits in Lewy bodies of PD. Journal of the Neurological Sciences. 2008; **266**:9-12

[76] Paleologou KE, Oueslati A, Shakked G, Rospigliosi CC, Kim HY, Lamberto GR, et al. Phosphorylation at S87 is enhanced in synucleinopathies, inhibits α-synuclein oligomerization, and influences synuclein-membrane interactions. Journal of Neuroscience. 2010;**30**:3184-3198

[77] Mbefo MK, Paleologou KE, Boucharaba A, Oueslati A, Schell H, Fournier M, et al. Phosphorylation of synucleins by members of the polo-like kinase family. Journal of Biological Chemistry. 2010;**285**:2807-2822

[78] Ng SS, Papadopoulou K, McInerny CJ. Regulation of gene expression and cell division by Polo-like kinases. Current Genetics. 2006;**50**:73-80

[79] Buck K, Landeck N, Ulusoy A, Majbour NK, El-Agnaf OM, Kirik D. Ser129 phosphorylation of endogenous α-synuclein induced by overexpression of polo-like kinases 2 and 3 in nigral dopamine neurons is not detrimental to their survival and function. Neurobiology of Disease. 2015;**30**(78):100-114

[80] Aubele DL, Hom RK, Adler M, Galemmo RA, Bowers S, et al. Selective and brain-permeable polo-like kinase-2 (Plk-2) inhibitors that reduce α-synuclein phosphorylation in rat brain. ChemMedChem. 2013;**8**(8):1295-1313

[81] Kauselmann G, Weiler M, Wulff P, Jessberger S, Konietzko U, et al. The polo-like protein kinases Fnk & Snk associate with a Ca^{2+} and integrin-binding protein and are regulated dynamically with synaptic plasticity. EMBO Journal. 1999;**18**:5528-5539

[82] Seeburg DP, Pak D, Sheng M. Polo-like kinases in the nervous system. Oncogene. 2005;**24**:292-298

[83] Qing H, Wong W, McGeer EG, McGeer PL. Lrrk2 phosphorylates α-synuclein at serine 129: PD implications. Biochemical and Biophysical Research Communications. 2009;**387**:149-152

[84] Guerreiro PS, Huang Y, Gysbers A, Cheng D, Gai WP, Outeiro TF, et al. LRRK2 interactions with α-synuclein in PD brain and in cell models. Journal of Molecular Medicine (Berlin). 2013;**91**:513-522

[85] Venda LL, Cragg SJ, Buchman VL, Wade-Martins R. α-Synuclein and dopamine at the crossroads of PD. Trends in Neurosciences. 2010;**33**(12):559-568

[86] Duda JE, Giasson BI, Mabon ME, Miller DC, Golbe LI, Lee VM, Trojanowski JQ. Concurrence of α-synuclein and tau brain pathology in the Contursi kindred. Acta Neuropathologica. 2002;**104**(1):7-11

[87] Giasson BI, Duda JE, Quinn SM, Zhang B, Trojanowski JQ, Lee VM. Neuronal α-synucleinopathy with severe movement disorder in mice expressing A53T human α-synuclein. Neuron. 2002;**34**(4):521-533

[88] Giasson BI, Covy JP, Bonini NM, Hurtig HI, Farrer MJ, et al. Biochemical & pathological characterization of Lrrk2. Annals of Neurology. 2006;**59**(2):315-322

[89] Greenbaum EA, Graves CL, Mishizen-Eberz AJ, Lupoli MA, Lynch DR, et al. The E46K mutation in α-synuclein increases amyloid fibril formation. Journal of Biological Chemistry. 2005;**280**(9):7800-7807

[90] Kosik KS, Orecchio LD, Binder L, Trojanowski JQ, Lee VY, Lee G. Epitopes that span the tau molecule are shared with paired helical filaments. Neuron.1988;**1**(9):817-825

[91] Otvos L, Feiner L, Lang E, Szendrei GI, Goedert M, Lee VM. Monoclonal antibody PHF-1 recognizes tau protein phosphorylated at serine residues 396 and 404. Journal of Neuroscience Research. 1994;**39**(6):669-673

[92] McFarland NR, Fan Z, Xu K, Schwarzschild MA, Feany MB, Hyman BT, McLean PJ. α-synuclein S129 phosphorylation mutants do not alter nigrostriatal toxicity in a rat model of PD. Journal of Neuropathology and Experimental Neurology.2009;**68**:515-524

[93] Lashuel HA, Overk CR, Oueslati A, Masliah E. The many faces of α-synuclein: From structure and toxicity to therapeutic target. Nature Reviews Neuroscience. 2013;**14**:38-48

[94] Lee VM, Trojanowski JQ. Mechanisms of PD linked to pathological α-synuclein: New targets for drug discovery. Neuron. 2006;**52**:33-38

[95] Mahul-Mellie AL, Fauvet B, Gysbers A, Dikiy I, Oueslati A, Georgeon S, et al. c-Abl phosphorylates α-synuclein & regulates its degradation, implication for α-synuclein clearance & contribution to the pathogenesis of PD. Human Molecular Genetics. 2014;**23**(11):2858-2879

[96] Oueslati A, Schneider BL, Aebischer P, Lashuel HA. Polo-like kinase 2 regulates selective autophagic α-synuclein clearance & suppresses its toxicity in vivo. Proceedings of the National academy of Sciences of the United States of America. 2013;**110**:E3945-E3954

[97] Chau KY, Ching HL, Schapira AH, and Cooper JM. Relationship between α-synuclein phosphorylation, proteasomal inhibition & cell death: Relevance to PD pathogenesis. Journal of Neurochemistry. 2009;**110**:1005-1013

[98] Yin G, Lopes da Fonseca T, Eisbach SE, Anduaga AM, Breda C, et al. α-Synuclein interacts with the switch region of Rab8a in a Ser129 phosphorylation-dependent manner. Neurobiology of Disease. 2014;**70**:149-161

[99] Hara S, Arawaka S, Sato H, Machiya Y, Cui C, Sasaki A, Koyama S, Kato T. Serine 129 phosphorylation of membrane-associated α-synuclein modulates dopamine transporter function in a G protein-coupled receptor kinase-dependent manner. Molecular Biology of the Cell. 2013;**24**:1649-1660, S1641-S1643.

[100] Yamada M, Iwatsubo T, Mizuno Y, Mochizuki H. Overexpression of α-synuclein in rat SNpc results in loss of dopaminergic neurons, phosphorylation of α-synuclein & activation of caspase-9: Resemblance to pathogenetic changes in PD. Journal of Neurochemistry. 2004;**91**:451-461

[101] Sampathu DM, Giasson BI, Pawlyk AC, Trojanowski JQ, Lee VM. Ubiquitination of α-synuclein is not required for formation of pathological inclusions in α-synucleinopathies. The American Journal of Pathology. 2003;**163**(1):91-100

[102] Buchman VL, Hunter HJ, Pinõn LG, Thompson J, Privalova EM, Ninkina NN, Davies AM. Persyn, a member of the synuclein family, has a distinct pattern of expression in the developing nervous system. Journal of Neuroscience. 1998;**18**(22):9335-9341

[103] Jensen PH, Hager H, Nielsen MS, Hojrup P, Gliemann J, & Jakes R. α-synuclein binds to Tau and stimulates the protein kinase A-catalyzed tau phosphorylation of serine residues 262 and 356. Journal of Biological Chemistry. 1999;**274**:25481-25489

[104] Bhaskar K, Yen SH, Lee G. Disease-related modifications in tau affect the interaction between Fyn & tau. Journal of Biological Chemistry. 2005;**280**:35119-35125

[105] Samuel F, Flavin WP, Iqbal S, Pacelli C, Renganathan SD, Trudeau LE, Campbell EM, Fraser PE, Tandon A. Effects of Serine 129 phosphorylation on α-synuclein aggregation, membrane association, and internalization. Journal of Biological Chemistry. 2016;**291**(9):4374-4385

[106] Sugeno N, Takeda A, Hasegawa T, Kobayashi M, Kikuchi A, Mori F, Wakabayashi K, Itoyama Y. Serine 129 phosphorylation of α-synuclein induces unfolded protein response mediated cell death. Journal of Biological Chemistry. 2008;**283**:23179-23188

[107] Saha AR, Hill J, Utton MA, Asuni AA, Ackerley S, Grierson AJ., et al. Parkinson's disease α-synuclein mutations exhibit defective axonal transport in cultured neurons. Journal of Cell Science. 2004;**117**:1017-1024

[108] Fernandez CO, Hoyer W, Zweckstetter M, Jares-Erijman EA, Subramaniam V, Griesinger C, et al. NMR of α-synuclein-polyamine complexes elucidates the mechanism and kinetics of induced aggregation. EMBO Journal. 2004;**23**:2039-2046

[109] Giasson BI, Forman MS, Higuchi M, Golbe LI, Graves CL, et al. Initiation and synergistic fibrillization of tau & α-synuclein. Science. 2003;**300**:636-640

[110] da Azeredo SS, Schneider BL, Cifuentes-Diaz C, et al. Phosphorylation does not prompt, nor prevent, the formation of α-synuclein toxic species in a rat model of PD. Human Molecular Genetics. 2009;**18**:872-887

[111] Gorbatyuk OS, Li S, Sullivan LF, Chen W, Kondrikova G, Manfredsson FP, Mandel RJ, Muzyczka N. The phosphorylation state of Ser-129 in human α-synuclein determines neurodegeneration in a rat model of PD. Proceedings of the National Academy of Sciences of the United States of America. 2008;**105**:763-768

[112] Ganapathy K, Datta I, Sowmithra S, Joshi P, and Bhonde R. Influence of 6-hydroxydopamine toxicity on α-Synuclein phosphorylation, resting vesicle expression, and vesicular dopamine release. Journal of Cellular Biochemistry. 2016;**9999**:1-18

[113] Gonçalves S, Outeiro TF. Assessing the subcellular dynamics of α-synuclein using photoactivation microscopy. Molecular Neurobiology. 2013;**47**:1081-1092

[114] Monti B, Gatta V, Piretti F, Raffaelli SS, Virgili M, Contestabile A. Valproic acid is neuroprotective in the rotenone rat model of PD: Involvement of α-synuclein. Neurotoxicity Research. 2010;**17**(2):130-141

[115] Siddiqui A, Chinta SJ, Mallajosyula JK, Rajagopolan S, Hanson I, et al. Selective binding of nuclear α-synuclein to the PGC1 α promoter under conditions of oxidative stress may contribute to losses in mitochondrial function: Implications for PD. Free Radical Biology and Medicine. 2012;**53**:993-1003

[116] Goers J, Manning-Bog AB, McCormack AL, Millett IS, Doniach S, Di Monte, DA, et al. Nuclear localization of α-synuclein and its interaction with histones. Biochemistry. 2003;**42**:8465-8471

[117] Kontopoulos E, Parvin JD, Feany MB. α-Synuclein acts in the nucleus to inhibit histone acetylation and promote neurotoxicity. Human Molecular Genetics. 2006;**15**:3012-3023

[118] Perfeito R, Lázaro DF, Outeiro TF, Rego AC. Linking α-synuclein phosphorylation to reactive oxygen species formation and mitochondrial dysfunction in SH-SY5Y cells. Molecular and Cellular Neuroscience. 2014;**62**:51-59

[119] Cuervo AM, Stefanis L, Fredenburg R, Lansbury PT, Sulzer D. Impaired degradation of mutant α-synuclein by chaperone-mediated autophagy. Science. 2004;**305**(5688): 1292-5

[120] Schel H, Hasegawa T, Neumann M, Kahle PJ. (2009) Nuclear & neuritic distribution of serine-129 phosphorylated α-synuclein in transgenic mice. Neuroscience. 2009;**160**:796-804

[121] Dyllick-Brenzinger M, D'Souza CA, Dahlmann B, Kloetzel PM, Tandon A. Reciprocal effects of α-synuclein overexpression and proteasome inhibition in neuronal cells and tissue. Neurotoxicity Research. 2010;**17**:215-227

[122] Febbraro F, Sahin G, Farran A, Soares S, Jensen PH, Kirik D, Romero-Ramos M. Ser129D mutant α-synuclein induces earlier motor dysfunction while S129A results in distinctive pathology in a rat model of PD. Neurobiology of Disease. 2013;**56**:47-58

[123] Oueslati A, Fournier M, Lashuel HA. Role of post-translational modifications in modulating the structure, function & toxicity of α-synuclein: Implications for PD pathogenesis & therapies. Progress in Brain Research. 2010;**183**:115-145

[124] Paleologou KE, Schmid AW, Rospigliosi CC, Kim HY, Lamberto GR, et al. Phosphorylation at Ser-129 but not the phosphomimics S129E/D inhibits the fibrillation of α-synuclein. Journal of Biological Chemistry. 2008;**283**:16895-16905

[125] Machiya Y, Hara S, Arawaka S, Fukushima S, Sato H, et al. Phosphorylated α-synuclein at Ser-129 is targeted to the proteasome pathway in a ubiquitin-independent manner. Journal of Biological Chemistry. 2010;**285**:40732-40744

[126] Gurevich EV, Tesmer JJG, Mushegian A, Gurevich VV. G-protein-coupled receptor kinases: More than just kinases and not only for GPCRs. Pharmacology & Therapeutics. 2012;**133**(1):40-69

[127] Walker DG, Lue LF, Adler CH, Shill HA, Caviness JN, et al. Arizona Parkinson Disease C. Changes in properties of serine 129 phosphorylated α-synuclein with progression of Lewy-type histopathology in human brains. Experimental Neurology. 2013;**240**:190-204

[128] Zhou J, Broe M, Huang Y, Anderson JP, Gai WP, et al. Changes in the solubility and phosphorylation of α-synuclein over the course of PD. Acta Neuropathologica. 2011;**121**:695-704

[129] Hong Z, Shi M, Chung KA, Quinn JF, Peskind ER, et al. DJ-1 and α-synuclein in human cerebrospinal fluid as biomarkers of PD. Brain. 2010;**133**(3):713-726

[130] Kang JH, Irwin DJ, Chen-Plotkin AS, Siderowf A, Caspell C, et al. Association of cerebrospinal fluid β-amyloid 1-42, T-tau, P-tau181, and α-synuclein levels with clinical features of drug-naive patients with early PD. JAMA Neurology. 2013;**70**(10):1277-1287

[131] Mollenhauer B, Cullen V, Kahn I, Krastins B, Outeiro TF, et al. Direct quantification of CSF α-synuclein by ELISA and first cross-sectional study in patients with neurodegeneration. Experimental Neurology. 2008;**213**(2):315-325

[132] Foulds PG, Diggle P, Mitchell JD, Parker A, Hasegawa M, et al. A longitudinal study on α-synuclein in blood plasma as a biomarker for PD. Scientific Reports. 2013;**3**:2540

[133] Stewart T, Sossi V, Aasly JO, Wszolek ZK, Uitti RJ, et al. Phosphorylated α-synuclein in PD: Correlation depends on disease severity. Acta Neuropathologica Communications. 2015;**3**(1):7

[134] Wang Y, Shi M, Chung KA, Zabetian CP, Leverenz JB, et al. Phosphorylated α-synuclein in PD. Science Translational Medicine. 2012;**4**(121):121

[135] Doppler K, Ebert S, Uceyler N, Trenkwalder C, Ebentheuer J, et al. Cutaneous neuropathy in PD: A window into brain pathology. Acta Neuropathologica. 2014;**128**:99-109

[136] Pouclet H, Lebouvier T, Coron E, Neunlist M, Derkinderen P. Lewy pathology in gastric and duodenal biopsies in PD. Movement Disorders. 2012;**27**:708

[137] Hilton D, Stephens M, Kirk L, Edwards P, Potter R, et al. Accumulation of α-synuclein in the bowel of patients in the pre-clinical phase of PD. Acta Neuropathologica. 2014;**127**:235-241

[138] Brown DR. Interactions between metals & α-synuclein function or artefact? FEBS Journal. 2007;**274**(15):3766-3774

[139] Hyun-Ju SH, Ju-Hyun LE, Chang CS, Jongsun KI. Copper (II)-induced self-oligomerization of α-synuclein. Biochemical Journal. 1999;**340**(3):821-828

[140] Paik SR, Shin HJ, Lee JH. Metal-catalyzed oxidation of α-synuclein in the presence of copper (II) & hydrogen peroxide. Archives of Biochemistry and Biophysics. 2000;**378**(2):269-277

Chapter 7

Ecto-Phosphorylation and Regeneration of the Adult Central Nervous System

Yoshinori Takei and Yoko Amagase

Additional information is available at the end of the chapter

http://dx.doi.org/10.5772/intechopen.70030

Abstract

Phosphorylation of ecto-domains of membrane proteins and extracellular matrix proteins, which is termed ecto-phosphorylation, activates intracellular signalling and has roles in several physiological processes including cell adhesion, fertilisation and fibrinolysis. We demonstrated that ecto-phosphorylation can promote endogenous neurogenesis in the damaged central nervous system (CNS), augmenting its functional recovery. Thus, regulation of ecto-phosphorylation could be a platform for development of therapeutic methods against CNS injury. Regeneration of the damaged CNS is long-awaited. While transplantation of neuronal progenitor cells is expected to be the first platform to develop the therapy, the potential of endogenous neurogenesis as a source of new neurons has been expected to be an inexpensive and non-invasive regenerative medicine for CNS injury. In this review, we focused on the spinal cord as a model of CNS recovery from traumatic injury. The spinal cord is the simplest part of the CNS and its function is well known. Therefore, estimation of recovery is easier than other part of the CNS. Firstly, we introduce endogenous neural stem cells (NSCs) in the adult spinal cord and their behaviour after injury and then discuss effects of ecto-phosphorylation, which induces regeneration of the adult spinal cord.

Keywords: adult neurogenesis, adult neural stem cells, activation of quiescent stem cells, spinal cord injury

1. Introduction

The adult mammalian central nervous system (CNS) loses its self-regeneration ability, whereas it contains neural stem cells (NSCs) that can differentiate into both neurons and glial cells [1, 2]. Under physiological conditions, neurogenesis in the adult CNS can be observed at the dentate gyrus of the hippocampus and the subventricular zone [3]. Neurogenesis is strictly inhibited in other regions of the adult CNS. After injury of the adult CNS, endogenous NSCs produce glial cells that are involved in formation of the glial scar. However, the stem cells rarely produce neuronal cells, which can contribute to regeneration of the neuronal network damaged by the injury. Thus, when the adult CNS is damaged, its neuronal network is not regenerated and incurable paralysis can be occurred. If cell fate decision of endogenous NSCs could be controlled in the injured CNS, the cells might be a source of new neurons and remyelinating oligodendrocytes to repair the damaged neuronal network. A recent report demonstrated that application of adenosine triphosphate (ATP) and a protein kinase promoted differentiation of neuronal cells in the damaged spinal cord, and diminished paralysis caused by the injury [4]. Thus, ecto-phosphorylation may provide a novel platform for regenerative medicine of the damaged CNS, in which endogenous NSCs are used for a source of new neurons.

2. Neural stem cells in the adult spinal cord

NSCs can be characterised by the ability of self-proliferation and of differentiation into both neuronal and glial cells [5]. Proliferating cells in the intact adult spinal cord are mostly oligodendrocyte precursors [6]. Other than that, ependymal cells existing around the central canal proliferate moderately, and radial glial cells existing throughout the spinal cord are also proliferative. These three types of cells are known to have potential to produce neurons at least *in vitro* and are proposed to be NSCs in the adult spinal cord [7–10].

When cultured NSCs derived from the spinal cord are transplanted into the adult spinal cord again, the differentiated cells from the transplanted NSCs are mainly astrocytes [11]. However, when the same NSCs are transplanted into a part of the hippocampus where neurons are generated through life, they produce neurons. This report demonstrates that NSCs derived from the adult spinal cord have ability to produce neurons, but the ability is inhibited by the microenvironment around NSCs in the spinal cord.

3. NSCs after spinal cord injury

The group of Frisen made transgenic mice, in which each type of stem cells was genetically labelled [6]. They observed cell fates of those stem cells after spinal cord injury. Radial glial cells produced only astrocytes after traumatic injury of the spinal cord, and oligodendrocyte precursors produced oligodendrocytes. Proliferation of ependymal cells was enhanced after spinal cord injury, and the cells moved to the injury site, producing astrocytes and very few oligodendrocytes (**Table 1**). Thus, endogenous neurogenesis is not activated after spinal cord injury.

http://dx.doi.org/10.5772/intechopen.70030

NSCs in the intact spinal cord	Genetic labelling	Protein markers	Increase at the lesion (fold)	Descendant cells (contribution)
Ependymal cells	FoxJ1 promoter	Sox9	4–5	Astrocytes (53%)
		Vimentin		Oligodendrocytes (3.2%)
Astrocytes	Connexin 30 promoter	Sox9	2	Astrocytes (47%)
		GFAP		
Oligodendrocytes	Olig2 promoter	Olig2	2	Oligodendrocytes (97%)
		Sox10		
		APC (mature oligodendrocytes)		

Table 1. Cell fate of NSCs after spinal cord injury (based on data from Ref. [6]).

4. Regulation of NSCs in the spinal cord

Whilst transplantation of cultured neuronal precursor cells is expected to provide a new therapy for spinal cord injury, attempts to use endogenous NSCs as a source of new neurons have been examined, as summarised in the sections.

4.1. Inflammatory cytokines

After spinal cord injury, pro-inflammatory cytokines including TNF-alpha, IL-6, and IL-1beta are secreted in the injured site, inducing inflammation. IL-6 promotes differentiation of NSCs into astrocytes. Suppression of IL-6 signalling decreases the production of astrocytes induced by spinal cord injury, promoting functional recovery of damaged spinal cord [12, 13].

4.2. Growth factor

Growth factors can modulate not only proliferation and differentiation of NSCs, but also survival, neurite outgrowth, and synapse plasticity of differentiated neurons. Both FGF2 and EGF suppress differentiation of NSCs, which in turn promoting proliferation. Administration of FGF2 and EGF into the lateral ventricle accelerates growth of not only subventricular zone NSCs but also NSCs around the central canal of spinal cord [14]. IGF-1 induces the production of oligodendrocytes by inhibiting bone morphogenetic protein (BMP) signals through induction of bone morphogenetic protein such as Smad6, Smad7 and Noggin [15]. When fibroblasts that can secrete Brain-derived neurotrophic factor (BDNF) are transplanted into the spinal cord injury lesion, oligodendrocytes production is induced [16]. After spinal cord injury, the exogenous delivery of nerve growth factor (NGF) can induce growth of corticospinal axons in rats [17, 18], whereas NT3 elicits growth of corticospinal axons [19, 20]. Glial cell-derived neurotrophic factor (GDNF) induces growth of motor and dorsal column sensory axons after partial and complete spinal cord transections and induces remyelination [21]. These reports suggest that control of growth factors in the injured area of the spinal cord may promote regeneration of the injured spinal cord. However, clinical trials using systemic delivery of growth factors for various disorders have failed either as a result of lack of efficiency or unacceptable side effects, or both [22, 23].

4.3. Transcription factor and growth factors

Neurogenin2 and Mash-1 are transcription factors required for neuronal differentiation. Production of neurons and oligodendrocytes at the lesion of spinal cord is induced by infection of retroviruses that express those transcription factors followed by application of BDNF, FGF2, and EGF [9]. However, it is unknown whether the treatment with transfection and growth factors can promote recovery of the function, or not.

5. Extracellular phosphorylation

Protein phosphorylation occurs not only in the intracellular space but also at the extracellular space (**Figure 1**). Phosphorylation of proteins located at the cell surface has been reported in many types of cells, including platelets [24, 25], monocytes [26], osteoblasts [27], vascular smooth

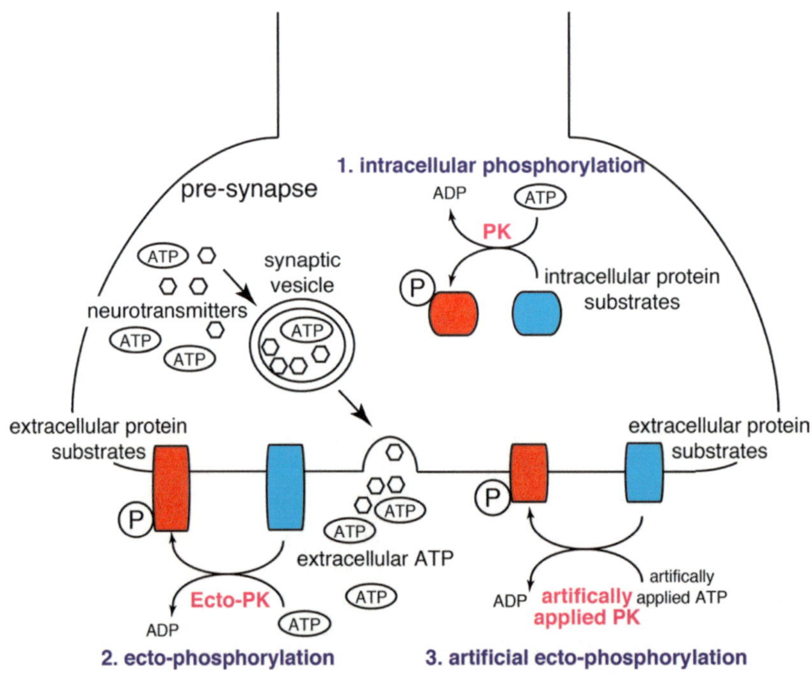

Figure 1. Three types of phosphorylation. (1) Intracellular phosphorylation by intracellular PKs. This type of phosphorylation has been studied extensively and intensively. (2) Endogenous ecto-phosphorylation by ecto-PKs. Some PKs can be secreted from cells through typical exsocytosis. Those ecto-PKs use extracellular ATP, which are secreted from many types of cells, including neurons and glial cells, as the donor of phosphorus residue for protein phosphorylation. (3) Artificial ecto-phosphorylation. In this case, both ATP and PKs are artificially applied. Any kinases that do not exist in the extracellular space can be applied, and they are applicable at any places even where endogenous PKs are not functional. Therefore, responses that endogenous ecto-PKs do not evoke can be expected.

muscle cells [28] and neurons [29]. Furthermore, some of extracellular matrix proteins are known to be phosphorylated [26, 30–32]. Yalak and Vogel identified 770 different phosphorylation sites in 66 extracellular proteins or in proteins with extracellular domain by annotation of secreted phosphorylated protein data available in public repositories [33]. Extracellular protein phosphorylation, which is termed ecto-phosphorylation, has been implicated in several physiological processes, including adhesion and migration of leukocyte and macrophage, fertilisation and fibrinolysis. Ecto-protein kinases (ecto-PKs) acting on the outer surface of the plasma membrane are reported to catalyse those phosphorylation. Ecto-PKs use extracellular ATP as a source of the phosphate group [28]. Thus far, FAM20C, protein kinase C (PKC), protein kinase A (PKA), casein kinase 1 (CK1) and casein kinase 2 (CK2) are reported as ecto-PKs [34, 35]. In addition to phosphorylation by ecto-PKs, ecto-domain of membrane proteins and proteins secreted from cells can be phosphorylated by kinases in the Golgi apparatus, which proteins pass through for secretion and for location to the plasma membrane. Ecto-domains of neuroglican C [36] and of amyloid beta precursor protein [37] can be phosphorylated by both ecto-PKs and kinases in the Golgi apparatus. Endogenous ecto-phosphorylations described in this chapter are summarised in **Table 2**.

These findings demonstrate that ecto-phosphorylation, as well as intracellular phosphorylation, can activate substrate proteins, inducing intracellular signalling.

Ecto-phosphorylation	Location	Effects	Cells
105, 39, 20 kDa proteins	Membrane	NGF-dependent neurite outgrowth	PC12 cells
12, 13 kDa proteins	Membrane	Correlation with neurite outgrowth	Primary embryonic chick neurons
48/50 kDa protein	Membrane	Long-term potentiation	Hippocampal pyramidal neurons
Laminin	Extracellular matrix	Cell adhesion and migration	–
Collagen XVII	Extracellular matrix	Unknown	–
Vitronectin	Extracellular matrix	Unknown	–
Fibronectin	Extracellular matrix	Unknown	–
NCAM	Membrane	Unknown	Neuronal cells
MAP1B	Post-synaptic area	Synapse formation	Cortical neurons
Beta-amyloid	Secreted	Promotion of aggregation	–

Table 2. Endogenous phosphorylation of ecto-domains of membrane proteins and extracellular matrix proteins described in the chapter.

6. Extracellular phosphorylation and neurite outgrowth machinery of neurons

In 1989, extracellular ATP was reported to stimulate uptake of noradrenaline into PC12 adrenal pheochromocytoma cells [38]. Addition of either ATP or ATPgammaS, but not adenosine diphosphate (ADP), guanosine tri-phosphate (GTP) or AppNHp, increased noradrenaline uptake into PC12 cells. This suggests that added ATP was used as the donor of phosphate group in the phosphorylation reaction. Although protein responsible for the regulation of noradrenaline uptake is still unknown, 105, 39 and 20 kDa proteins at the surface of PC12 cells were identified as substrates for ecto-phosphorylation [39]. NGF, which can induce neuronal differentiation and neurite outgrowth of PC12 cells, enhances extracellular phosphorylation of PC12 cells [39]. A non-permeable and non-specific inhibitor of kinases, K252b, blocked both NGF-dependent neurite outgrowth and enhancement of extracellular phosphorylation [39, 40].

In 1995, primary embryonic chick neurons were incubated with radioactive ATP to show ecto-protein kinase activity [41]. Incorporation of radioactivity was detected with 116, 105, 67, 53, 17, 13 and 12 kDa proteins. Addition of either a pseudo-substrate peptide for PKC or a neutralising antibody against catalytic domain of PKC inhibited phosphorylation of 67, 13 and 12 kDa proteins, indicating that ecto-PKC catalysed phosphorylation of those proteins. Extracellular phosphorylation of 12 and 13 kDa proteins by PKC is regulated by development of the brain and is correlated with neurite outgrowth of neurons in the CNS. The 12 and 13 kDa proteins have not been identified.

7. Long-term potentiation and extracellular phosphorylation

Long-term potentiation is the long-lasting improvement in neuronal communication, which is the major cellular mechanism for learning and memory [42, 43]. Extracellular phosphorylation of neuronal surface proteins is implicated in long-term potentiation in the hippocampus [44–46]. Using mouse hippocampal slices, addition of either ATP or ATPgammaS, but not AppNHp, was reported to amplify permanently the magnitude of the population spike [44]. A 48/50 kDa protein at the surface of hippocampal pyramidal neurons becomes phosphorylated during the amplification [45]. Addition of a neutralising antibody against PKC inhibits the extracellular 48/50 kDa protein phosphorylation, and it blocks the stabilisation of long-term potentiation in hippocampal slices [45]. Thus, PKC-mediated extracellular phosphorylation is required for maintenance of hippocampal long-term potentiation. However, the 48/50 kDa protein has not been identified.

8. Extracellular phosphorylation and synapse

Synapse formation is an essential step to make neuronal network. K-252b, which is a non-permeable and non-specific inhibitor of kinases, inhibits synapse formation between cortical neurons *in vitro* [47]. This suggests that ecto-kinase sensitive to K-252b has a role in synapse formation. Lately, the same group has indicated that MAP1B, which is a tubulin-binding protein distributed in axon, especially in growth corn, is a substrate of ecto-protein kinases, and K-252b inhibits phosphorylation of MAP1B [48]. Originally, MAP1B was thought to be located in the cytoplasm. However, at least some splice valiant forms of MAP1B can be located at the plasma membrane [49]. The membrane-bound MAP1B is located in the post-synaptic area, but not in pre-synaptic area [50]. The membrane-bound MAP1B can interact with myelin-associated glycoprotein (MAG) and the binding enhances phosphorylation of MAP1B [51]. Neither the kinase catalysing MAP1B phosphorylation nor the phosphorylation site in MAP1B has been elucidated. These reports suggest a possibility that ecto-phosphorylation of membrane-bound MAP1B has a role in regulation of synapse formation.

These findings suggest that ecto-phosphorylation can regulate neurite outgrowth, long-term potentiation and synapse formation of neurons. This gives an idea that ecto-phosphorylation may promote regeneration of the damaged neuronal network.

9. Ecto-domain phosphorylation of NgR

9.1. Effects on terminally differentiated neurons

NgR is a receptor of myelin-associated glycoproteins, Nogo-A, MAG and oligodendrocyte-myelin glycoprotein (OMgp) [52, 53]. Binding of those glycoproteins to NgR inhibits axonal outgrowth of neurons, at least *in vitro*. NgR signals activate intracellular Rho protein regulating rearrangement of cytoskeleton, which suppresses axonal outgrowth and synapse formation. Paralysis by spinal cord injury can be reduced by administration of an antibody against Nogo-A protein [54]. Genetic depletion of NgR shows NgR is partially responsible for limiting the regeneration of certain fibre systems in the adult CNS [53, 55].

We found that the extracellular domain of NgR can be phosphorylated by PKA and CK2, and that the phosphorylation inhibits binding of NgR agonists [56]. Interestingly, both kinases phosphorylate the same amino acid residue of the ecto-domain of NgR (**Figure 2**). *In vitro* study indicated that ecto-domain phosphorylation of NgR by either PKA or CK2 can overcome the inhibition of axonal outgrowth by NgR agonists Nogo-A, MAG and OMgp.

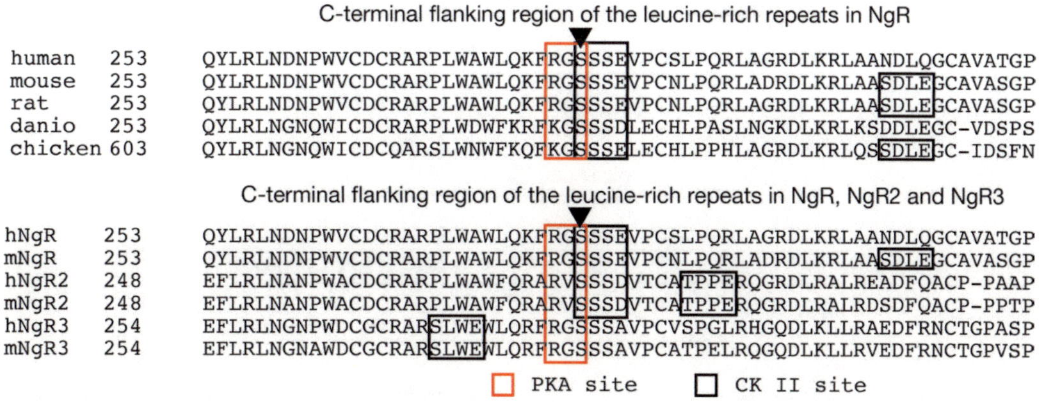

Figure 2. Phosphorylation sites in the ecto-domains of NgRs. (▼) indicates the phosphorylation site detected by mass spectrometer after the *in vitro* phosphorylation assay.

9.2. Effects on differentiation of NSCs

We reported that neurogenesis in the injured spinal cord can be augmented by artificial ecto-phosphorylation with continuous application of active PKA and ATP [4]. Since the treatment promoted functional recovery of the injured spinal cord, ecto-phosphorylation can be a novel target of therapies against paralysis caused by spinal cord injury.

In vitro assay indicated that NgR is expressed in ependymal cell-like NSCs derived from adult mouse spinal cord [4]. Noteworthy, proliferation of ependymal cells is enhanced after spinal cord injury, and the cells move to the injury site [6]. When differentiation of the NSCs was induced in the presence of NgR inhibitors, a transient increase of cells expressing a transcription factor Olig2 was observed on day 5 of *in vitro* differentiation. Olig2 can suppress differentiation into astrocytes but can promote differentiation into oligodendrocytes and motor neurons. Whereas oligodendrocytes were not observed in the descendant cells on day 14 of *in vitro* differentiation, increased proportion of neuronal cells was observed. When active PKA and ATP were applied on mice with spinal cord injury, NgR within the spinal cord was phosphorylated, and cells expressing neuronal precursor cell markers, such as doublecortin and neurogenin2, were increased. Significant improvement of hindlimbs' paralysis was also observed.

Our report proposed that myelin proteins released from damaged oligodendrocytes can suppress both axonal outgrowth from survived neurons and neurogenesis of NSCs, through

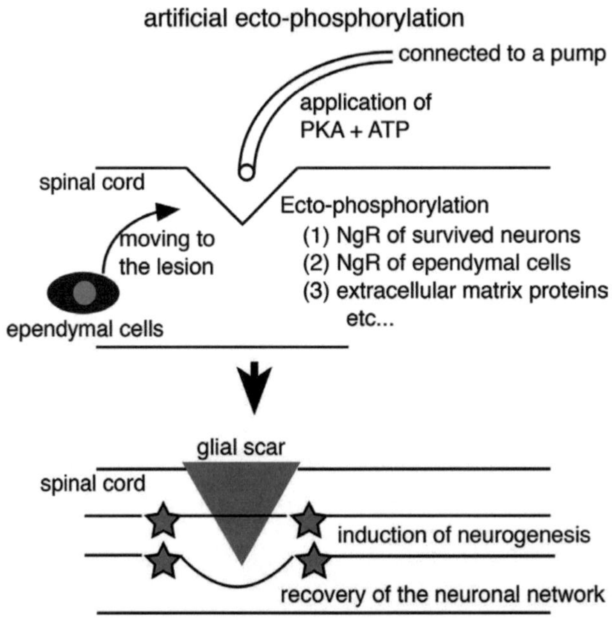

Figure 3. Possible effects of artificial ecto-phosphorylation induced by direct application of active kinases. Application of active PKA and ATP induces ecto-phosphorylation of NgR expressed in both differentiated neurons and NSCs and of other proteins. The phosphorylation can promote neurogenesis, enhancing functional recovery of the spinal cord from traumatic injury.

NgR activation. However, our results do not eliminate corporation of ecto-phosphorylation of other proteins. Application of active PKA and ATP to the injured spinal cord should phosphorylate many ecto-domains of membrane proteins and extracellular matrix proteins, in addition to NgR. The favourable effects on recovery from spinal cord injury could result from cooperation of ecto-phosphorylation of those proteins (**Figure 3**).

10. Phosphorylation of cell adhesion molecules and extracellular matrix proteins

Ecto-phosphorylation of proteins described below was originally reported with non-neuronal cells or in the extracellular matrix. However, when active PKA and ATP were applied to the damaged spinal cord, phosphorylation of these extracellular matrix proteins can be expected and similar effects should be induced.

10.1. Laminin

Laminin is an extracellular matrix protein found in basement membranes. Laminin interacts with cell surface proteins, influencing not only cell attachment on the basement membrane but also cell function and differentiation. Laminin is phosphorylated by CK2 [26]. This phosphorylation enhances heparin binding to laminin, cell attachment and migration. Laminin can be phosphorylated by PKC, in addition to CK2. Phosphorylation by PKC enhances self-assembly, heparin binding and cell attachment [57].

Laminin has important roles in neuronal differentiation of human embryonic stem cells [58], proliferation of human neural stem cells [59], netrin-mediated axonal guidance [60] and NGF-mediated neurite outgrowth from both PC-12 cells [61] and primary cultured neurons derived from mouse dorsal root ganglia [62]. Thus, phosphorylation of laminin could contribute to regulation of neuronal differentiation, migration and neurite outgrowth.

10.2. Collagen XVII

Collagen XVII can bind to alpha6 integrin, contributing to cell adhesion and motility. The serine 544 in the extracellular domain of collagen XVII is phosphorylated by CK2 and the phosphorylation inhibits shedding of the extracellular domain by metalloproteases of the A disintegrin and metalloproteinase (ADAM) family [32]. Although the exact function of collagen XVII is unknown, collagen XVII is expressed in the CNS and its distribution is changed in neurodegenerative disorders [63].

10.3. Vitronectin

Vitronectin is a glycoprotein in blood and extracellular matrix. Threonine 50 and 57 of vitronectin can be phosphorylated by CK2 [31] and the phosphorylation enhances its binding to both alpha(v)beta3 integrin [64] and urokinase receptor [65]. Furthermore, the serine 378 of vitronectin is phosphorylated by PKA, which induces a conformational change and enhances the phosphorylation by CK2 [66]. Vitronectin is known to regulate differentiation of cerebellar granule cell precursor cells [67].

In addition, fibronectin [30] and neuronal cell adhesion molecules [29] are known to be phosphorylated. However, effects of their phosphorylation are unknown.

11. Perspective

It has been expected to develop a CNS injury treatment via activation of endogenous stem cells, because it may provide simple and inexpensive therapy with minimal invasion. We demonstrated that application of PKA and ATP can induce neurogenesis of endogenous NSCs in the damaged spinal cord, diminishing paralysis caused by the damage. The ecto-domain of NgR in the spinal cord is phosphorylated by the PKA application. NgR expression is detected in ependymal cell-like NSCs derived from the spinal cord, and ecto-phosphorylation of NgR promotes neuronal differentiation. However, multiple sites of extracellular proteins and domains should be simultaneously phosphorylated by the application. The favourable effects of the application are possibly due to a cooperation of phosphorylation of those proteins, including ecto-phosphorylation of NgR. Artificial ecto-phosphorylation could be the platform for development of therapies for cure of paralysis caused by spinal cord injury. More study is required for revealing the precise mechanism of which artificial ecto-phosphorylation promotes regeneration of the damaged CNS.

Acknowledgements

This work was supported by the Grant-In Aid for Scientific Research (C) (JSPS KAKENHI number 23500392) from the Japan Society for the Promotion of Science and by the Seamless Technology Transfer Program through Target-Driven R&D (AS221Z01416G and AS232Z00373F) of the Japan Science and Technology Agency. This work was also partially supported by the Shimizu Foundation for Immunology and Neuroscience.

Author details

Yoshinori Takei[1]* and Yoko Amagase[2]

*Address all correspondence to: ytakei@pharm.kyoto-u.ac.jp

1 Department of Nanobio Drug Discovery, Graduate School of Pharmaceutical Science, Kyoto University, Kyoto, Japan

2 Department of Pathophysiology, Faculty of Pharmaceutical Sciences, Doshisha Women's College of Liberal Arts, Kodo, Kyotanabe City, Kyoto, Japan

References

[1] Gross CG. Neurogenesis in the adult brain: Death of a dogma. Nature Reviews Neuroscience. 2000;**1**:67-73. DOI: 10.1038/35036235

[2] Okano H, Sawamoto K. Neural stem cells: Involvement in adult neurogenesis and CNS repair. Philosophical Transactions of the Royal Society of London Series B, Biological Sciences. 2008;**363**:2111-2122. DOI: 10.1098/rstb.2008.2264

[3] Ming GL, Song H. Adult neurogenesis in the mammalian brain: Significant answers and significant questions. Neuron. 2011;**70**:687-702. DOI: 10.1016/j.neuron.2011.05.001

[4] Suehiro K, Nakamura Y, Xu S, Uda Y, Matsumura T, Yamaguchi Y, Okamura H, Yamashita T, Takei Y. Ecto-domain phosphorylation promotes functional recovery from spinal cord injury. Scientific Reports. 2014;**4**:4972. DOI: 10.1038/srep04972

[5] Bottai D, Fiocco R, Gelain F, Defilippis L, Galli R, Gritti A, Vescovi LA. Neural stem cells in the adult nervous system. Journal of Hematotherapy & Stem Cell Research. 2003;**12**:655-670. DOI: 10.1089/152581603607326877

[6] Barnabe-Heider F, Goritz C, Sabelstrom H, Takebayashi H, Pfrieger FW, Meletis K, Frisen J. Origin of new glial cells in intact and injured adult spinal cord. Cell Stem Cell. 2010;**7**:470-482. DOI: 10.1016/j.stem.2010.07.014

[7] Johansson CB, Momma S, Clarke DL, Risling M, Lendahl U, Frisen J. Identification of a neural stem cell in the adult mammalian central nervous system. Cell. 1999;**96**:25-34

[8] Horner PJ, Power AE, Kempermann G, Kuhn HG, Palmer TD, Winkler J, Thal LJ, Gage FH. Proliferation and differentiation of progenitor cells throughout the intact adult rat spinal cord. The Journal of Neuroscience. 2000;**20**:2218-2228

[9] Ohori Y, Yamamoto S, Nagao M, Sugimori M, Yamamoto N, Nakamura K, Nakafuku M. Growth factor treatment and genetic manipulation stimulate neurogenesis and oligodendrogenesis by endogenous neural progenitors in the injured adult spinal cord. The Journal of Neuroscience. 2006;**26**:11948-11960. DOI: 10.1523/JNEUROSCI.3127-06.2006

[10] Meletis K, Barnabe-Heider F, Carlen M, Evergren E, Tomilin N, Shupliakov O, Frisen J. Spinal cord injury reveals multilineage differentiation of ependymal cells. PLoS Biology. 2008;**6**:e182. DOI: 10.1371/journal.pbio.0060182

[11] Shihabuddin LS, Horner PJ, Ray J, Gage FH. Adult spinal cord stem cells generate neurons after transplantation in the adult dentate gyrus. The Journal of Neuroscience. 2000;**20**:8727-8735

[12] Okada S, Nakamura M, Mikami Y, Shimazaki T, Mihara M, Ohsugi Y, Iwamoto Y, Yoshizaki K, Kishimoto T, Toyama Y, Okano H. Blockade of interleukin-6 receptor suppresses reactive astrogliosis and ameliorates functional recovery in experimental spinal cord injury. Journal of Neuroscience Research. 2004;**76**:265-276. DOI: 10.1002/jnr.20044

[13] Okano H, Okada S, Nakamura M, Toyama Y. Neural stem cells and regeneration of injured spinal cord. Kidney International. 2005;**68**:1927-1931. DOI: 10.1111/j.1523-1755.2005.00621.x

[14] Kojima A, Tator CH. Intrathecal administration of epidermal growth factor and fibroblast growth factor 2 promotes ependymal proliferation and functional recovery after spinal cord injury in adult rats. Journal of Neurotrauma. 2002;**19**:223-238. DOI: 10.1089/08977150252806974

[15] Hsieh J, Aimone JB, Kaspar BK, Kuwabara T, Nakashima K, Gage FH. IGF-I instructs multipotent adult neural progenitor cells to become oligodendrocytes. Journal of Cell Biology. 2004;**164**:111-122. DOI: 10.1083/jcb.200308101

[16] McTigue DM, Horner PJ, Stokes BT, Gage FH. Neurotrophin-3 and brain-derived neurotrophic factor induce oligodendrocyte proliferation and myelination of regenerating axons in the contused adult rat spinal cord. The Journal of Neuroscience. 1998;**18**:5354-5365

[17] Tuszynski MH, Gabriel K, Gage FH, Suhr S, Meyer S, Rosetti A. Nerve growth factor delivery by gene transfer induces differential outgrowth of sensory, motor, and noradrenergic neurites after adult spinal cord injury. Experimental Neurology. 1996;**137**:157-173. DOI: 10.1006/exnr.1996.0016

[18] Grill RJ, Blesch A, Tuszynski MH. Robust growth of chronically injured spinal cord axons induced by grafts of genetically modified NGF-secreting cells. Experimental Neurology. 1997;**148**:444-452. DOI: 10.1006/exnr.1997.6704

[19] Grill R, Murai K, Blesch A, Gage FH, Tuszynski MH. Cellular delivery of neurotrophin-3 promotes corticospinal axonal growth and partial functional recovery after spinal cord injury. The Journal of Neuroscience. 1997;**17**:5560-5572

[20] Tuszynski MH, Grill R, Jones LL, Brant A, Blesch A, Löw K, Lacroix S, Lu P. NT-3 gene delivery elicits growth of chronically injured corticospinal axons and modestly improves functional deficits after chronic scar resection. Experimental Neurology. 2003;**181**:47-56

[21] Blesch A, Tuszynski MH. Cellular GDNF delivery promotes growth of motor and dorsal column sensory axons after partial and complete spinal cord transections and induces remyelination. Journal of Comparative Neurology. 2003;**467**:403-417. DOI: 10.1002/cne.10934

[22] Apfel SC. Neurotrophic factor therapy—Prospects and problems. Clinical Chemistry and Laboratory Medicine. 2001;**39**:351-355. DOI: 10.1515/CCLM.2001.055

[23] Apfel SC. Nerve growth factor for the treatment of diabetic neuropathy: What went wrong, what went right, and what does the future hold. International Review of Neurobiology. 2002;**50**:393-413

[24] Hatmi M, Gavaret JM, Elalamy I, Vargaftig BB, Jacquemin C. Evidence for cAMP-dependent platelet ectoprotein kinase activity that phosphorylates platelet glycoprotein IV (CD36). Journal of Biological Chemistry. 1996;**271**:24776-24780

[25] Babinska A, Hogan MV, Sobocki T, Sobocka MB, Ehrlich YH, Kornecki E. Identification of ecto-PKC on surface of human platelets: Role in maintenance of latent fibrinogen receptors. The American Journal of Physiology-Heart and Circulatory Physiology. 2000;**278**:H2008-H2019

[26] Trachana V, Christophorides E, Kouzi-Koliakos K, Koliakos G. Laminin-1 is phosphorylated by ecto-protein kinases of monocytes. The International Journal of Biochemistry & Cell Biology. 2005;**37**:478-492. DOI: 10.1016/j.biocel.2004.08.001

[27] Zhu X, Luo C, Ferrier JM, Sodek J. Evidence of ectokinase-mediated phosphorylation of osteopontin and bone sialoprotein by osteoblasts during bone formation in vitro. Biochemical Journal. 1997;**323**:637-643

[28] Dumler I, Stepanova V, Jerke U, Mayboroda OA, Vogel F, Bouvet P, Tkachuk V, Haller H, Gulba DC. Urokinase-induced mitogenesis is mediated by casein kinase 2 and nucleolin. Current Biology. 1999;**9**:1468-1476

[29] Ehrlich YH, Davis TB, Bock E, Kornecki E, Lenox RH. Ecto-protein kinase activity on the external surface of neural cells. Nature. 1986;**320**:67-70. DOI: 10.1038/320067a0

[30] Imada S, Sugiyama Y, Imada M. Fibronectin phosphorylation by ecto-protein kinase. Experimental Cell Research. 1988;**179**:554-564

[31] Seger D, Gechtman Z, Shaltiel S. Phosphorylation of vitronectin by casein kinase II. Identification of the sites and their promotion of cell adhesion and spreading. Journal of Biological Chemistry. 1998;**273**:24805-24813

[32] Zimina EP, Fritsch A, Schermer B, Bakulina AY, Bashkurov M, Benzing T, Bruckner-Tuderman L. Extracellular phosphorylation of collagen XVII by ecto-casein kinase 2 inhibits ectodomain shedding. Journal of Biological Chemistry. 2007;**282**:22737-22746. DOI: 10.1074/jbc.M701937200

[33] Yalak G, Vogel V. Extracellular phosphorylation and phosphorylated proteins: Not just curiosities but physiologically important. Science Signaling. 2012;**5**:re7. DOI: 10.1126/scisignal.2003273

[34] Redegeld FA, Caldwell CC, Sitkovsky MV. Ecto-protein kinases: Ecto-domain phosphorylation as a novel target for pharmacological manipulation? Trends in Pharmacological Sciences. 1999;**20**:453-459

[35] Yalak G, Ehrlich YH, Olsen BR. Ecto-protein kinases and phosphatases: An emerging field for translational medicine. Journal of Translational Medicine. 2014;**12**:165. DOI: 10.1186/1479-5876-12-165

[36] Yamauchi S, Tokita Y, Aono S, Matsui F, Shuo T, Ito H, Kato K, Kasahara K, Oohira A. Phosphorylation of neuroglycan C, a brain-specific transmembrane chondroitin sulfate proteoglycan, and its localization in the lipid rafts. Journal of Biological Chemistry. 2002;**277**:20583-20590. DOI: 10.1074/jbc.M200909200

[37] Walter J, Capell A, Hung AY, Langen H, Schnolzer M, Thinakaran G, Sisodia SS, Selkoe DJ, Haass C. Ectodomain phosphorylation of beta-amyloid precursor protein at two distinct cellular locations. Journal of Biological Chemistry. 1997;**272**:1896-1903

[38] Hardwick JC, Ehrlich YH, Hendley ED. Extracellular ATP stimulates norepinephrine uptake in PC12 cells. Journal of Neurochemistry. 1989;**53**:1512-1518

[39] Pawlowska Z, Hogan MV, Kornecki E, Ehrlich YH. Ecto-protein kinase and surface protein phosphorylation in PC12 cells: Interactions with nerve growth factor. Journal of Neurochemistry. 1993;**60**:678-686

[40] Rogers MV, Buensuceso C, Montague F, Mahadevan L. Vanadate stimulates differentiation and neurite outgrowth in rat pheochromocytoma PC12 cells and neurite extension in human neuroblastoma SH-SY5Y cells. Neuroscience. 1994;**60**:479-494

[41] Hogan MV, Pawlowska Z, Yang HA, Kornecki E, Ehrlich YH. Surface phosphorylation by ecto-protein kinase C in brain neurons: A target for Alzheimer's beta-amyloid peptides. Journal of Neurochemistry. 1995;**65**:2022-2030

[42] Cooke SF, Bliss TV. Plasticity in the human central nervous system. Brain. 2006;**129**:1659-1673. DOI: 10.1093/brain/awl082

[43] Neves G, Cooke SF, Bliss TV. Synaptic plasticity, memory and the hippocampus: A neural network approach to causality. Nature Reviews Neuroscience. 2008;**9**:65-75. DOI: 10.1038/nrn2303

[44] Wieraszko A, Ehrlich YH. On the role of extracellular ATP in the induction of long-term potentiation in the hippocampus. Journal of Neurochemistry. 1994;**63**:1731-1738

[45] Chen W, Wieraszko A, Hogan MV, Yang HA, Kornecki E, Ehrlich YH. Surface protein phosphorylation by ecto-protein kinase is required for the maintenance of hippocampal long-term potentiation. Proceedings of the National Academy of Sciences of the United States of America. 1996;**93**:8688-8693

[46] Fujii S, Kuroda Y, Ito K, Kato H. Long-term potentiation induction—A synaptic catch mechanism released by extracellular phosphorylation. Neuroscience. 2000;**96**:259-266

[47] Kuroda Y, Ichikawa M, Muramoto K, Kobayashi K, Matsuda Y, Ogura A, Kudo Y. Block of synapse formation between cerebral cortical neurons by a protein kinase inhibitor. Neuroscience Letters. 1992;**135**:255-258

[48] Muramoto K, Taniguchi H, Kawahara M, Kobayashi K, Nonomura Y, Kuroda Y. A substrate of ecto-protein kinase is microtubule-associated protein 1B in cortical cell cultures undergoing synaptogenesis. Biochemical and Biophysical Research Communications. 1994;**205**:1467-1473. DOI: 10.1006/bbrc.1994.2830

[49] Tanner SL, Franzen R, Jaffe H, Quarles RH. Evidence for expression of some microtubule-associated protein 1B in neurons as a plasma membrane glycoprotein. Journal of Neurochemistry. 2000;**75**:553-562

[50] Kawakami S, Muramoto K, Ichikawa M, Kuroda Y. Localization of microtubule-associated protein (MAP) 1B in the postsynaptic densities of the rat cerebral cortex. Cellular and Molecular Neurobiology. 2003;**23**:887-894

[51] Franzen R, Tanner SL, Dashiell SM, Rottkamp CA, Hammer JA, Quarles RH. Microtubule-associated protein 1B: A neuronal binding partner for myelin-associated glycoprotein. Journal of Cell Biology. 2001;**155**:893-898. DOI: 10.1083/jcb.200108137

[52] Pernet V, Schwab ME. The role of Nogo-A in axonal plasticity, regrowth and repair. Cell and Tissue Research. 2012;**349**:97-104. DOI: 10.1007/s00441-012-1432-6

[53] Schwab ME, Strittmatter SM. Nogo limits neural plasticity and recovery from injury. Current Opinion in Neurobiology. 2014;**27**:53-60. DOI: 10.1016/j.conb.2014.02.011

[54] Schwab ME. Nogo and axon regeneration. Current Opinion in Neurobiology. 2004;**14**:118-124. DOI: 10.1016/j.conb.2004.01.004

[55] Kim JE, Liu BP, Park JH, Strittmatter SM. Nogo-66 receptor prevents raphespinal and rubrospinal axon regeneration and limits functional recovery from spinal cord injury. Neuron. 2004;**44**:439-451. DOI: 10.1016/j.neuron.2004.10.015

[56] Takei Y. Phosphorylation of nogo receptors suppresses nogo signaling, allowing neurite regeneration. Science Signaling. 2009;**2**:ra14. DOI: 10.1126/scisignal.2000062

[57] Koliakos G, Trachana V, Gaitatzi M, Dimitriadou A. Phosphorylation of laminin-1 by protein kinase C. Molecules and Cells. 2001;**11**:179-185

[58] Ma W, Tavakoli T, Derby E, Serebryakova Y, Rao MS, Mattson MP. Cell-extracellular matrix interactions regulate neural differentiation of human embryonic stem cells. BMC Developmental Biology. 2008;**8**:90. DOI: 10.1186/1471-213X-8-90

[59] Hall PE, Lathia JD, Caldwell MA, Ffrench-Constant C. Laminin enhances the growth of human neural stem cells in defined culture media. BMC Neuroscience. 2008;**9**:71. DOI: 10.1186/1471-2202-9-71

[60] Ratcliffe EM, D'Autreaux F, Gershon MD. Laminin terminates the netrin/DCC mediated attraction of vagal sensory axons. Developmental Neurobiology. 2008;**68**:960-971. DOI: 10.1002/dneu.20634

[61] Rankin SL, Guy CS, Mearow KM. Neurite outgrowth is enhanced by laminin-mediated down-regulation of the low affinity neurotrophin receptor, p75NTR. Journal of Neurochemistry. 2008;**107**:799-813. DOI: 10.1111/j.1471-4159.2008.05663.x

[62] Plantman S, Patarroyo M, Fried K, Domogatskaya A, Tryggvason K, Hammarberg H, Cullheim S. Integrin-laminin interactions controlling neurite outgrowth from adult DRG neurons in vitro. Molecular and Cellular Neuroscience. 2008;**39**:50-62. DOI: 10.1016/j.mcn.2008.05.015

[63] Seppanen A, Suuronen T, Hofmann SC, Majamaa K, Alafuzoff I. Distribution of collagen XVII in the human brain. Brain Research. 2007;**1158**:50-56. DOI: 10.1016/j.brainres.2007.04.073

[64] Seger D, Seger R, Shaltiel S. The CK2 phosphorylation of vitronectin. Promotion of cell adhesion via the alpha(v)beta 3-phosphatidylinositol 3-kinase pathway. Journal of Biological Chemistry. 2001;**276**:16998-17006. DOI: 10.1074/jbc.M003766200

[65] Stepanova V, Jerke U, Sagach V, Lindschau C, Dietz R, Haller H, Dumler I. Urokinase-dependent human vascular smooth muscle cell adhesion requires selective vitronectin phosphorylation by ectoprotein kinase CK2. Journal of Biological Chemistry. 2002;**277**:10265-10272. DOI: 10.1074/jbc.M109057200

[66] Schvartz I, Kreizman T, Brumfeld V, Gechtman Z, Seger D, Shaltiel S. The PKA phosphorylation of vitronectin: Effect on conformation and function. Archives of Biochemistry and Biophysics. 2002;**397**:246-252. DOI: 10.1006/abbi.2001.2699

[67] Wechsler-Reya RJ. Caught in the matrix: How vitronectin controls neuronal differentiation. Trends in Neurosciences. 2001;**24**:680-682

Section 3

Tools to Study Kinases

Chapter 8

Computational Modeling of Complex Protein Activity Networks

Stefano Schivo, Jeroen Leijten,
Marcel Karperien and Janine N. Post

Additional information is available at the end of the chapter

http://dx.doi.org/10.5772/intechopen.69804

Abstract

Because of the numerous entities interacting, the complexity of the networks that regulate cell fate makes it impossible to analyze and understand them using the human brain alone. Computational modeling is a powerful method to unravel complex systems. We recently described the development of a user-friendly computational tool, Analysis of Networks with Interactive MOdeling (ANIMO). ANIMO is a powerful tool to formalize knowledge on molecular interactions. This formalization entails giving a precise mathematical (formal) description of molecular states and of interactions between molecules. Such a model can be simulated, thereby in silico mimicking the processes that take place in the cell. In sharp contrast to classical graphical representations of molecular interaction networks, formal models allow in silico experiments and functional analysis of the dynamic behavior of the network. In addition, ANIMO was developed specifically for use by biologists who have little or no prior modeling experience. In this chapter, we guide the reader through the ANIMO workflow using osteoarthritis (OA) as a case study. WNT, IL-1β, and BMP signaling and cross talk are used as a concrete and illustrative model.

Keywords: WNT, IL1β, BMP, cartilage, computational model, ANIMO, cell signaling, network modeling

1. Introduction

1.1. Signal transduction networks

At any given point in time, cells are exposed to many different signals from their environment. Cells will have to interpret this multitude of signals they receive. Signal transduction networks relay and integrate signals from membrane-bound receptors, via protein activation, to the nucleus in order to regulate cellular processes such as gene transcription, metabolism, proliferation, differentiation, and apoptosis (programmed cell death).

Kinases play a key role in signal transduction by transferring phosphate groups to their substrates in a process called phosphorylation [1–4]. In this context, phosphorylation is basically a way to hand over a signal. In practice, kinases function by phosphorylating serine, threonine, or tyrosine residues on downstream substrates, thereby inducing conformational changes and/or charge alterations, resulting in modulation of protein activities [5].

Signal transduction pathways are connected to other signal transduction pathways in a mechanism we call cross talk. Due to this cross talk, signaling pathways are part of extensive signaling networks. Ultimately, dynamic changes in the signaling network determine cell fate. Insight into this network-regulating cell fate is important for controlling stem cell differentiation, understanding diseases such as cancer and osteoarthritis (OA), defining better diagnostics based on biomarker expression, and designing precision therapies.

1.2. Network topology and dynamics

To understand signaling networks, graphical representations are very useful (and widely used). In such graphical network representations, network topology and protein interactions are displayed in a static way. This is very useful for understanding network topology but fails to show the dynamics of the network interactions. In addition, as networks become large with many interactions between signaling molecules, it becomes harder to comprehend and predict the speed of the network interactions. Since we want to understand the dynamics of signaling networks, we need to incorporate quantitative aspects like activity levels and the timing of interactions. Understanding the interplay between the quantitative dynamics and the distributed and concurrent nature of networks with large numbers of components is a formidable task; this task can only be successfully undertaken by using methods and techniques that are adequately supported by software tools.

1.3. Computational modeling of signaling networks

The systems biology approach to understanding biological systems starts off from a scientific question and then follows an empirical cycle—or rather a positive spiral—of knowledge/theory → model → hypotheses → experiments → observations → update and/or refinement of knowledge/theory, until an answer to the original question is found (**Figure 1**). The model plays a pivotal role in this cycle:

1. To organize data and store knowledge
2. To structure reasoning and discussion
3. To perform in silico experiments and derive hypotheses

An in silico model is always a simplified representation of biological reality and is never the aim in itself. Rather, it is a powerful means in the process of gaining an understanding of the

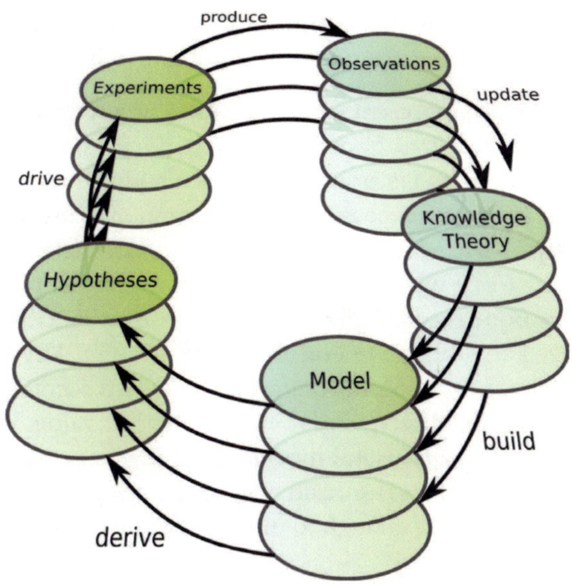

Figure 1. The empirical spiral: applying the empirical cycle in successive rounds leads to a gradual buildup of knowledge.

biological system. Given its role in the empirical cycle, the process of modeling is especially effective when applied by the experts with respect to a certain biological system. Biologists usually have a good sense of cause-and-effect relationships of molecular interactions. In addition, they are the most knowledgeable on the network topology and the dynamics of the biological system they are studying. Since they also benefit most from the generation of hypotheses and from an efficient experimental design, biologists would be the primary candidates to construct models of their research topic.

As models are a formalization of knowledge or theories, an underlying formalism is needed to express this knowledge. Different formal methods have been successfully applied to construct representations of biological systems. Among these methods are Boolean logic [6, 7], ordinary differential equations (ODEs, reviewed in Ref. [8]), interacting state machines [9, 10], process calculi [11, 12], timed automata [13–15], and Petri nets [16, 17]. Most of these formal methods have been implemented into software tools to aid the process of modeling.

Mastery of most existing modeling tools requires training and experience in mathematical modeling. In this respect, a lack of tradition in quantitative reasoning and formal methods within the biological community at large is still a stumbling block for widespread application of modeling of biological systems. To overcome this, we built an intuitive method for the construction of formal in silico models of the dynamics of molecular networks, supported by a user-friendly modeling tool, (Analysis of Networks with Interactive MOdeling (ANIMO) [18]).

1.4. ANIMO

ANIMO is an activity network tool, built as a plug-in to the network visualization program Cytoscape [19] and founded on the formalism of timed automata [13–15], but does not require the user to have any previous training in formal methods [20]. This provides the advantages of formal models (in silico experiments and model checking) without renouncing to usability.

Nodes in an ANIMO network represent an activity level of any given biological entity, e.g., proteins directly involved in signal transduction (e.g., kinases, growth factors, cytokines, genes, and mRNA. An *activity level* is associated to each node, to represent, for example, the relative amount of phosphorylated kinase or the concentration of mRNA. The activity level of a node can be altered by *interactions* with other nodes. ANIMO networks can include activations (→) and inhibitions (⊣), which will increase (resp. decrease) the activity level of the target node if the source node is active. For example, A → B will increase the activity level of B if A is active. The speed at which an interaction occurs is defined by its *k* parameter, which can be estimated qualitatively by choosing among a predefined set of options (*very slow*, *slow*, *medium*, *fast* and *very fast*) or by directly inputting a numerical value. We note that using the qualitative choices already leads to useful models: choosing, for example, a *slow* interaction to represents the production of a protein, and a *fast* one for a posttranslational modification such as phosphorylation is already enough to provide a realistic behavior in a network with the proper node topology [18, 20, 21].

A finer control on the network dynamics can be obtained by choosing for each interaction an approximated scenario which allows to describe the interaction. A choice is available among three scenarios:

- Scenario 1 (default): the interaction rate is linearly dependent on the *k* parameter and the activity level of the upstream node. This is the simplest scenario and is advised for all interactions when first building an ANIMO model.
- Scenario 2: the interaction rate depends on the *k* parameter and on the activity levels of both nodes. In particular, it is linearly dependent on the activity level of the upstream node and inversely dependent on the activity of the downstream node. This scenario is used to model reactions where the availability of substrate is a limiting factor.
- Scenario 3 (*AND* gate): the interaction rate depends on the *k* parameter and on the activity level of two user-defined nodes. The user can determine whether the dependency on a node's activity is linear or inverse. This scenario can be used to represent Boolean AND gates, such as "A *AND* B → C," where it is required that both A and B are active in order for C to become active.

Additionally, the *k* parameter can be manually set to numerical values, expanding the default qualitative choices. Methods for parameter fitting are also present in ANIMO, which allow to automatically adapt the parameters to a given data set [22]. These features are useful when comparing a model to experimental data and allow to easily try different parameter settings before needing to extend a model with new nodes or interactions.

1.5. Experimental requirements

Biological events can often be interpreted as changes in activity. Activities could be defined as changes in concentration, phosphorylation, or localization of a protein or changes in gene expression because they are causal factors with respect to downstream effects. Therefore, the state or concentration of the molecules can be described in terms of an activity. The more active the molecule is, the stronger it will affect downstream processes.

Experimental design could be performed according to these guidelines:

1. A negative control (t = 0) is needed to determine background activity levels.
2. Including a positive control for each of the measurements indicates the potential maximum intensity in the biological system. This allows scaling of activity between 0 and 100% to construct a nondimensional model, omitting the need for precise intracellular concentrations.
3. Include measurements of molecules that either have downstream effects in the model or can be used as an output of the model.
4. Include overlapping treatment conditions to normalize experimental data between different days or assay batches.
5. Include multiple time points to provide insight into the dynamic behavior of the system. To decide how many time points should be measured and what the optimum time range is, consider the following factors. Ideally, measurements are obtained at time points starting from t = 0 until the system reaches a steady state. For most primary effects in signal transduction networks, this means measuring more time points in the first 2–30 minutes after stimulation. When peak dynamics are expected, three time points are the absolute minimum to describe each peak, one before the peak, one as close as possible to the actual peak, and one after the peak. Five time points and more allow finding and describing a peak in more detail, especially in the presence of experimental noise. If no peak dynamics are expected, at least four time points should be measured. Try to avoid having the highest measurement value as the first or last value in your time series, as it will lead to uncertainty about the actual behavior of the system.

We can discern primary (or direct) effects or higher order (or indirect) effects after treatment. Indirect effects are those in which feedback is involved. For signal transduction, the primary effect occurs in time points up to 240/480 minutes. For gene expression, primary effects typically take 4–12 hours. Higher order effects occur at different time ranges, e.g., signal transduction could occur up to 24/48 hours; for gene expression involving higher order effects, for example, in the case of cell differentiation, effects can take up to several weeks.

2. Case study: ANIMO modeling of inflammatory signals in osteoarthritis

Many diseases are multifactorial, affected by many factors including genetic predisposition, age, trauma, sex, etc. These factors influence the network topology as well as its dynamics. This is hard to capture in static networks. To guide the reader through the ANIMO workflow, we use osteoarthritis as a case study.

Osteoarthritis (OA) is a painful, disabling disease with a high prevalence, occurring in about 15% of the population. The lifetime risk of knee OA is over 40% for man and almost 50% for women (reviewed in Ref. [23]). The lack of insight into the intricate signaling network of the cartilage has prevented the identification of highly needed disease-modifying osteoarthritic drugs (DMOADs). We aim to solve this by generating a comprehensive computational model of the signaling network in the cartilage [24]. In this chapter, we describe three important pathways in cartilage and OA development as a case study.

2.1. Osteoarthritis

Articular cartilage (AC) is a highly resilient tissue that covers the surfaces at the ends of long bones and ensures the pain-free and supple movement of our joints. The cartilage is mainly composed of one single cell type, the chondrocyte, which secretes and shapes the cartilaginous matrix that is necessary for its load-bearing properties. The biomechanical properties of the cartilage are dependent mainly on the composition, as well as the integrity of its matrix [25]. Once damaged, articular cartilage has low self-repair and regenerative capabilities eventually resulting in OA. This is due to its avascular nature, lack of innervation, and the embedding of chondrocytes in a dense matrix preventing cell migration. In addition, abnormalities in the cartilage-specific matrix cause a variety of skeletal malformation syndromes as well as adult-onset degenerative disorders such as OA.

OA is the most common form of arthritis and a leading cause of mobility-associated disability. OA is characterized by degeneration of articular cartilage, typical bone changes, and signs of inflammation, particularly in end-stage disease. The current management of OA is symptomatic, aimed at reduction of pain and at the end stage of disease total joint replacement as a successful treatment option for large joints (e.g., knee and hip) [26]. These treatments, however, do not cure the disease. There are no systemic drugs that can modify the disease process. Once the cartilage is damaged, no treatment exists that can intervene effectively, and the affected joint enters a disease continuum toward osteoarthritis.

Cartilage tissue homeostasis depends on a fine balance between catabolic (breakdown) and anabolic (buildup) processes. Homeostasis is regulated by a number of signaling pathways, including BMP and WNT signaling [27–31]. The amplitude of the signaling can be fine-tuned via antagonists in the extracellular space (reviewed in Ref. [32]). Typical catabolic pathways include the inflammatory pathways, including TNFα and IL-1β.

2.2. Osteoarthritis at the molecular level

OA is a disease caused by loss of homeostasis, resulting in altered mechanical and biochemical signals. Some of the key biochemical signals are growth factors such as WNT, IL-1β, transforming growth factor beta, bone morphogenetic proteins (BMPs), and Indian hedgehog homolog [27, 33, 34]. Mechanical stress on the extracellular matrix (ECM) plays a key role in OA development [27, 33, 34]. Any changes in this complex biological system, such as those caused by injury or aging, can disrupt cartilage homeostasis and lead to either catabolism characterized by expression of, for example, matrix metalloproteases (MMPs), and aggrecanases (ADAMs), or anabolism characterized by expression of collagen II and aggrecans [33].

As OA progresses the chondrocytes start to lose their characteristic phenotype, which in some cases results in differentiating into hypertrophic chondrocytes [35], resulting in endochondral ossification, by destroying the surrounding collagen II and replacing it with collagen X. Eventually, the hypertrophic chondrocytes will recruit osteoblasts that will proceed to form an osteophyte [35]. It is important to note that OA is not a disease that will damage the whole joint evenly. Throughout the cartilage there will be cells in different stages of differentiation, ranging from seemingly healthy cells to osteophyte forming hypertrophic chondrocytes.

The direct control of chondrogenic differentiation and hypertrophy is believed to be tightly regulated by the transcriptional activity of two main transcription factors: RUNX2, a transcription factor important for the regulation of hypertrophic differentiation, and SOX9, master transcription factor for chondrogenic development [36, 37]. The exact activity of these factors seems a key in determining the outcome of the chondrocyte phenotype.

The first steps in any computational modeling workflow are to thoroughly investigate the signal transduction pathways involved in the disease and to choose which pathways will be focused on. In this example, we will show the BMP and WNT signaling pathways for their importance in cartilage development and IL-1β as an inflammatory signal involved in osteoarthritis.

2.2.1. WNT signaling in the cartilage and osteoarthritis

The canonical Wnt pathway is crucial for cell survival and OA activation. The canonical pathway is characterized by the axin/glycogen synthase kinase-3β (GSK3-β) destruction complex which maintains low intracellular concentration of the key transcriptional regulator, β-catenin [33]. WNTs bind to the Frizzled (Fz) transmembrane receptors, resulting in the recruitment of the transmembrane protein LRP5/6. This complexation leads to the phosphorylation and dissociation of the destruction complex allowing β-catenin to accumulate and translocate to the nucleus [33]. In turn, β-catenin downregulates both collagen type 2A (*COL2A*) and SRY-box 9 (*SOX9*), leading to cell dedifferentiation and proliferation [38]. The WNT pathway can be activated by IL-1β through the phosphoinositide 3-kinase (PI3K) – protein kinase B (Akt) pathway [39, 40].

2.2.2. BMP2 signaling in osteoarthritis

BMP2 signaling is key pathway in the development of both the bone and cartilage. In endochondral bone formation, it is responsible for the clustering of the mesenchymal stem cells, the acquisition of the chondrocyte phenotype, and the final differentiation into hypertrophic chondrocytes [41, 42]. This final step is stopped in order to produce adult chondrocytes [41]. BMP2 is found in both healthy and OA adult chondrocytes [42]. BMP2 signaling occurs when BMP2 binds to its type 1 and type 2 receptors, which in turn phosphorylate mothers against decapentaplegic (SMAD) homologs 1, 5, and 8. Subsequently, SMAD 1, 5, or 8 dimers could bind to the ubiquitous SMAD-4 transcription factor [41]. BMP2 signaling can upregulate *Col2a*, *SOX9*, *ColX*, and *MMP13* gene expression [38, 41]. Once OA is advanced, BMP2 can cause hypertrophic differentiation of chondrocytes, leading to osteophyte formation [41].

2.2.3. Interleukin 1β signaling in osteoarthritis

IL-1β is a key pro-inflammatory cytokine that drives OA progression by inducing the expression of cartilage degrading enzymes such as matrix metalloproteinases (MMPs) [43, 44].

IL-1β signals by binding to the transmembrane IL-1 receptor I (IL-1RI), leading to the activation of multiple signaling pathways. The canonical IL-1β pathway signals through NF-κB, but IL-1β can also activate the p38-MAPK and JNK-MAPK pathways. The activated receptor then assembles two signaling proteins, myeloid differentiation primary response gene 88 (MYD88) and interleukin-1 receptor-activated protein kinase 4 (IRAK4). Together, the proteins form a stable IL-1-induced first signaling module and activate IRAK1 and IRAK2, which in turn activate TRAF6, PELI 1-3, TAK1, and MEKK3 [45]. IRAK1 also activates the inhibitor of nuclear factor B kinase (IKK) complex, which is necessary for the translocation of NF-κB (nuclear factor kappa-light-chain-enhancer of activated B cells) subunits to the core. The IKK complex consists out of IKK1 and IKK2 plus the regulatory subunit NF-κB essential modifier (NEMO) and phosphorylates IkB, the inhibitor of NF-κB, leading to its degradation [45, 46]. Due to the degradation of IkB, two NF-κB subunits, p50 and p65, are released and translocate into the nucleus, where they bind to conserved DNA motifs, which exist in many IL-1β responsive genes, including the genes for IL-1B [45, 46] and IκBα [45].

For the p38 MAPK and the JNK pathways, TAK1 and MEKK3 are mainly responsible, which activate MAPK kinase kinases (MKKs) 3, 4, 6, and 7 [45]. In addition, ERK1/ERK2 is also activated as a result of the activation of the IKK complex, which influences MKK1. All three MAPK pathways influence the activation protein 1 (AP-1), affecting the DNA expression of IL-1β response genes [45].

An increasing amount of data indicates the influence of IL-1β on the degeneration of extracellular matrix in the pathology of OA [47]. In addition, we have previously shown that WNT/β-catenin inhibits IL-1β-induced MMP expression in human articular cartilage [48]. Moreover, we showed that the WNT/β-catenin-regulated transcription factor TCF4 can bind to NF-κB, thereby enhancing NF-κB activity [30].

2.3. Defining an a priori network

The aim of computational modeling is not to provide the most complete representation of all the interactions in a signaling network, but to use as many interactions as needed to provide insight into cellular mechanisms. As such, models are indeed simplified representations of the real situation: one can choose a level of abstraction depending on the available information and the research question that is asked. The level of abstraction is always a trade-off between precision and feasibility.

In the case study presented here, we do not aim to build a precise model, but aim to show how building a model enables researchers of all levels of experience to visualize, summarize, and formalize models. Generating a relatively simple model in which key interactions are shown enables researchers to test and discuss various hypotheses quickly. With the obtained insight, one can then choose to validate only those hypotheses that the researchers will expect to truly yield new information. The model is then used as a backbone for the smart design of wet lab experiments rather than the trial-and-error methods that are traditionally used in the field.

2.3.1. Defining a research question

We base our research question on our analysis of the role of the individual pathways and their possible cross talk on OA development. Although the roles of WNT and BMP in cartilage and OA development are well described, the precise interactions between WNT, BMP, and IL-1β in regulating OA and chondrocyte hypertrophy are not fully understood (reviewed in Ref. [31]).

The model can be used for rapid and reiterative queries to derive and probe hypotheses such as whether IL-1β could influence cartilage homeostasis by modulating the activity of the cartilage and bone transcription factors SOX9 and RUNX2 and their downstream targets. Similarly, the role of BMP and WNT signalings, two important pathways in cartilage development and maintenance, can be explored on their potential modulatory roles on IL-1β expression and function.

To explore this example and provide a concrete guide through the ANIMO workflow, we will first draw a priori knowledge network, then formalize this network based on literature and our own data, and then perform a few simple in silico experiments that will be validated in the wet lab.

2.3.2. Drawing a priori knowledge network

To build an a priori network, we use KEGG (www.genome.jp/kegg) and WikiPathways (wikipathways.org) to decide on the topology of the proteins in the network.

We first draw an a priori network diagram that includes some of the most important factors in the signaling pathways of interest: IL-1β (based on: WikiPathways WP2637), WNT (e.g., the WNT homepage [49]), and BMP (e.g., WP1425). We include those intracellular molecules that we can actually measure in our experiments (see Sections 1.5 and 2.6), and added dashed lines to indicate other molecules important for the pathway were omitted (see Section 2.2).

In the WNT pathway, the inhibition of the destruction pathway leads to the inhibition of destruction of β-catenin, resulting in its upregulation and nuclear translocation [32]. So the net effect of WNT binding to its receptor is the increase of β-catenin activity. To simplify the model, we omit the many steps involving the double inhibition mechanisms in the WNT pathway, resulting in the simplified path WNT → WNTR → GSK3 → β-catenin → TCF/LEF.

SOX9 regulates expression of the matrix proteins collagen 2 and aggrecan. RUNX2 regulates transcription of collagens I and X and MMP13 [50, 51]. Since the activity of SOX9 and RUNX2 is key to the switch from the cartilage to hypertrophic cartilage, we included SOX9 and RUNX2 and some of their target genes in our diagram (**Figure 2**, **3A**).

2.4. Adding dynamics to the network

Once a priori knowledge network has been chosen, it needs to be drawn in ANIMO as a collection of nodes and interactions. For each node, a maximum number of activity levels can be chosen: unless a model is extremely large, it is safe to use 100 levels for all nodes. After providing the node with a name and an initial activity (which describes the state of the node at the start of a simulation), a description can optionally be added. Descriptions can be used as rationale for the presence of a certain node in the model, for example, a node description can contain citations to literature or references to own experimental results.

When adding an interaction between two nodes, a choice for an approximation scenario and a *k* parameter is necessary. The *k*-constants in our a priori knowledge network are not taken from literature as the strength of all interactions is assumed to be equal. This is a pragmatic decision, as many actual *k*-values are not described for most of the protein interactions in our network. In our initial models, we assume that there are in general two types of reactions: fast reactions for posttranslational modifications, such as phosphorylation, and slow reactions where gene transcription occurs. We therefore add reactions between nodes using these two types of reaction speed with a "Scenario 1" setting. We also add auto-inhibition to indicate inhibition as described in the literature for, e.g., receptor internalization, phosphatase activity, and, in the case of NF-κB, nuclear export as regulated by IκB. A more in-depth discussion on parameters, scenario choices, and network topology can be found in Refs. [21, 22].

Based on our experience, we expect proteins directly downstream of an activated receptor, such as p38 and JNK that are downstream of IL-1R, to be most activated by phosphorylation about 15–30 minutes after stimulation, and that the activity would be decreased to the starting situation between 60 and 240 minutes after stimulation. We therefore set the initial parameters to match these assumptions.

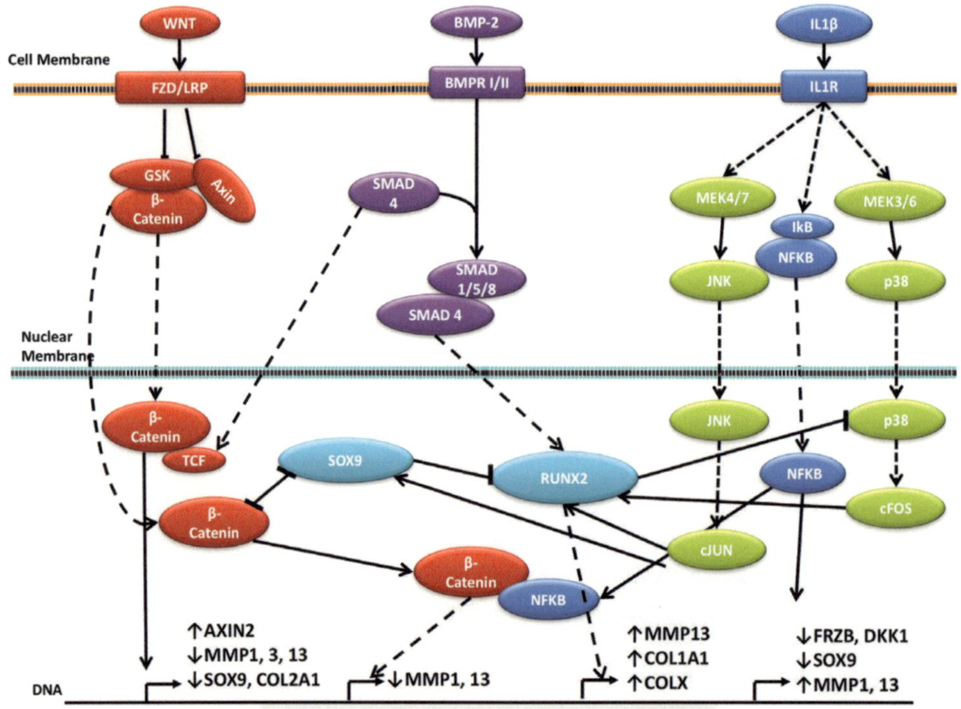

Figure 2. A priori knowledge network of WNT, BMP, and IL-1β pathways. IL-1β canonical (blue) and noncanonical signaling (green) showing cross talk with the transcription factors RUNX and SOX9 (both light blue), transcription factors from the WNT (red), and BMP2 (purple) pathways. Solid arrows indicate direct protein interaction, and dashed arrows indicate that intermediate protein interaction is omitted because no cross talk occurs between these proteins with other pathways. The colors indicate the canonical and noncanonical pathways corresponding to the external signals WNT, BMP2, and IL-1β.

2.5. Testing network effects of different stimuli in silico

Testing the effect of different stimuli in a computational network is performed in small steps. During each step the parameters in the model can be updated so that the dynamics of the various nodes in the network match our knowledge.

Step 1. What is the normal "steady-state" (=No Input) situation of the nodes represented in the network? For example, for articular chondrocytes it is known that the transcription factor SOX9 is active and that collagen 2 and aggrecan are expressed. It is also known that the WNT and IL-1β pathways are inactive and that BMP is active at a low level [52, 53]. We therefore can adjust our starting activities to these settings. This is generated in **Figure 4**. We display the activities of the proteins of which we plan to measure the phosphorylation, ERK1/2, GSK3, JNK, and p38 as well as the gene expression of AXIN2 and COL2A1. Initially, COL2A is expressed, indicating SOX9 activity.

Step 2. In a first test of the response of cells to various stimuli, we tested the presence of WNT starting from our "steady-state" model generated as described in Step 1. Since we do not starve our cells in the experiment, BMP2 is active at a low level of 20 activity units. After WNT addition we see that GSK3 becomes activated, and the activity peaks between 30 and 60 minutes after WNT addition and then trails off around 400 minutes. We see that AXIN2 becomes present between 2 and 4 hours after WNT treatment. This is probably faster than what can be expected from a newly synthesized mRNA. At the same time, due to the inactivation of SOX9 by β-catenin [54], we observe a reduction in the COL2A1 expression around 2–3 hours after WNT addition (not shown).

Step 3. We then tested the presence of IL-1β starting from our "steady-state" model. Again, BMP is active at a level of 20. We now observe that within 15 minutes of IL-1β addition, the three downstream kinases ERK1/2, p38, and JNK become active. In turn, these kinases activate RUNX2 [55], thereby activating its target genes. Due to the negative feedback loop from RUNX2 to JNK and p38, the activity curve is more narrow for these proteins than it is for ERK1/ERK2 [56]. GSK3 becomes slightly active via AKT activation by IL-1β. We see no reduction of COL2A mRNA expression and a transient activation of COL1 and COLX mRNA expression.

Step 4. Next, we want to see the effect of dual stimulation of WNT and IL-1β when starting from a healthy situation as described under Step 1. Addition of WNT and IL-1β, in the presence of 20% BMP, decreases the activity of SOX9 and therefore causes loss of COL2A1 expression. At the same time, RUNX2 becomes active, thereby activating MMP13, COL1, and COLX, ultimately leading to a hypertrophic phenotype.

Step 5. The next question was what is the effect of IL-1β and WNT in the presence of high BMP activity? In our model, the presence of high BMP activity is enough to prevent the loss of COL2A expression (not shown), while at the same time leading to induction of RUNX2 activity and the corresponding COL1 and COLX expression. However, we would have expected that the high levels of WNT and IL-1β would lead to reduced SOX9 activity as seen in articles describing OA (reviewed in Ref. [31]). We therefore need to adapt our network to match the literature data.

Step 6. It has been described that SOX9 and β-catenin influence each other's activity [54]. Also, RUNX2 suppresses SOX9 in bone formation [57], while SOX9 suppresses RUNX2 activity [36]. We added mutual inhibitions between SOX9 and RUNX2. This showed that when RUNX2 becomes active, it suppresses the SOX9-induced COL2 expression.

These few in silico experiments provide insight into the possible cross talk between these three pathways and their effect on SOX9 and RUNX2 activity and the possible effects of upstream signals. While asking the questions, we modified the parameters of the reactions, added inhibitory loops, and checked signaling cross talk in order to match the reaction speed in the model with our literature data or own experience.

This initial model now allows us to investigate the mechanism of IL-1β in inducing cellular hypertrophy by directly regulating SOX9 and RUNX2 activity. We hypothesize that IL-1β will increase expression of hypertrophic genes by upregulating RUNX2 activity and downregulating SOX9 activity (**Figure 3**).

2.6. Testing hypothesis by wet lab experiments

2.6.1. Designing experiment

The outcomes of the in silico experiments are used as guideline for the experimental setup for the wet lab validation. In this case, we questioned the effect of IL-1β on WNT signaling in the presence or absence of BMP signaling in the cartilage. For this we stimulated cells with IL-1β, WNT3A, or BMP2 either alone or in combinations. Parts of the data used in this chapter are published previously [21]. The other raw data can be obtained upon request.

2.6.2. Wet lab data

After the creation of the initial model with defined nodes, the next step was to obtain the data to fit into the model. This step was carried out mainly with wet lab data complemented with literature data. It is important that the analyses of wet lab data show consistency with well-known osteoarthritic cellular responses; these analyses are done prior to inclusion of experimental data into the model.

Figure 4 shows the measured and predicted activity of the various proteins of our network. We observe attenuations in, for example, p38, JNK, and ERK phosphorylation, where WNT partially inhibited the effect of IL-β on the phosphorylation of these proteins. We have already described these data [21]. When we add BMP to the cells, in combination with IL-1β and WNT, we see that in addition to the lower activity of the proteins, there is also a delay in the time by which the maximum activity is reached. This results in a delay and a reduction of the level of gene expression of all genes tested.

2.6.3. Comparing wet lab data to in silico data

The wet lab data obtained from the experiment were normalized and rescaled from 0 to 100 in order to be comparable with ANIMO's simulation results. For the complete normalization

http://dx.doi.org/10.5772/intechopen.69804

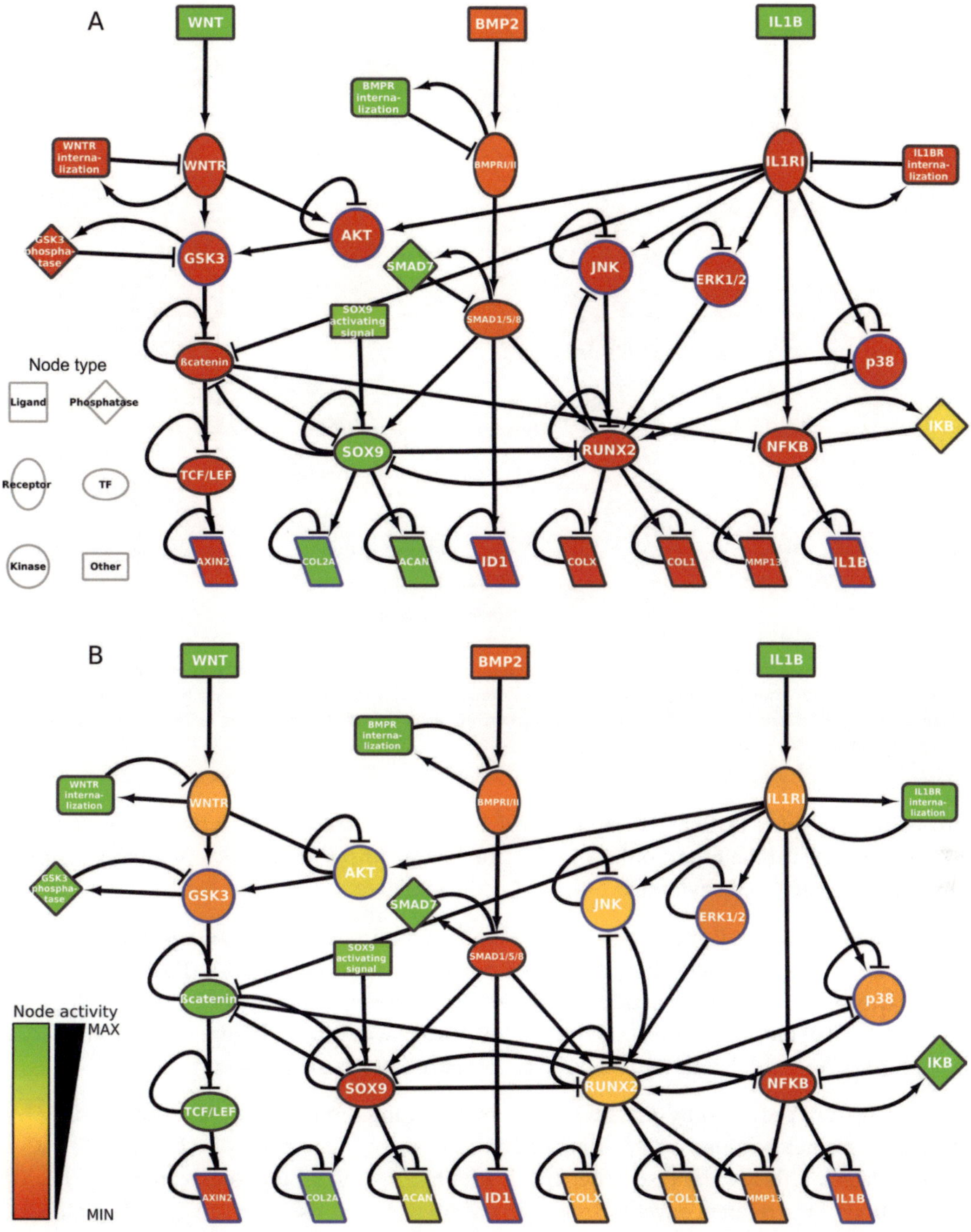

Figure 3. The ANIMO model built to represent the cross talk between the WNT, BMP, and IL-1β pathways. (A) The initial configuration of the model. (B) An example of simulation in ANIMO: activity levels of all nodes are shown after 120 minutes of treatment with WNT + IL-1β using color coding. The node colors are indicative of their activity at the indicated time points, with green being most active, via yellow to red, which is inactive as indicated in the figure, bottom left.

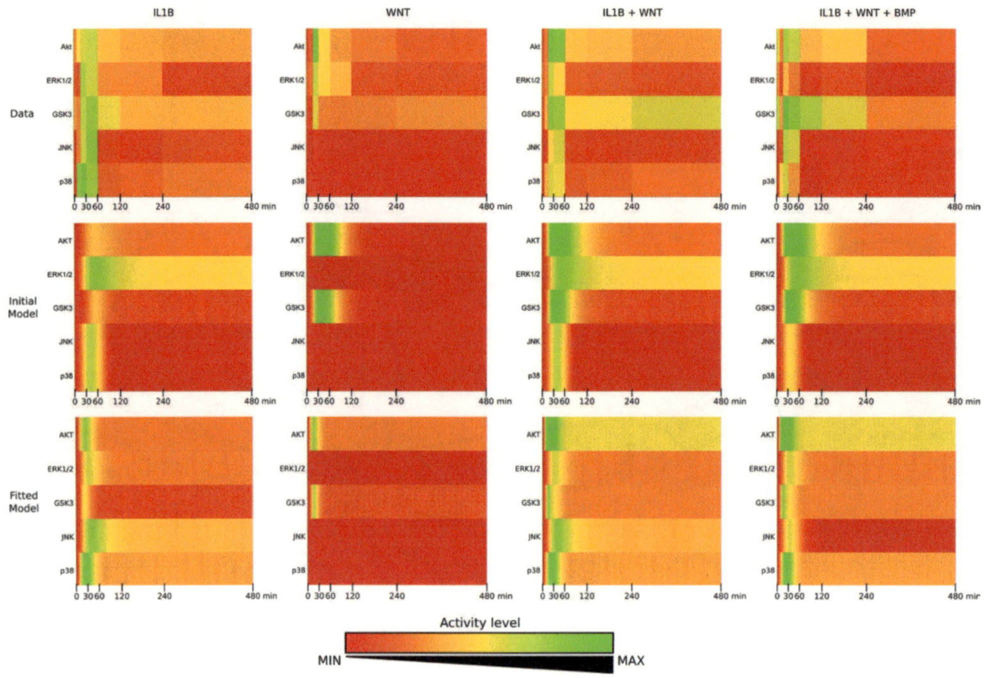

Figure 4. Comparing ANIMO's results with wet lab data: signal transduction. The results from two versions of the model are shown: the initial version with qualitative parameters (initial model) and the one obtained with ANIMO's automatic parameter fitting feature (fitted model). Colors are indicative of activity with green being most active and red, via yellow, inactive (see Legend).

procedure, we refer to Ref. [20] and ANIMO's manual. The resulting.csv tables, together with the model, can be found online, in the link below: http://fmt.cs.utwente.nl/tools/animo/content/models/Phosphorylation.

To evaluate the accuracy of the model, we compared its simulation results against the wet lab data concentrating at first on the signal transduction part of the network (**Figure 4**, first two rows). The initial match was already quite close, even if the parameters used in the model were all of qualitative nature (**Table 1**). The heat map graphs we show in **Figure 4** were obtained in ANIMO, where wet lab data can be directly compared to simulation results.

Another feature provided by ANIMO is parameter fitting, which allows to automatically try different parameter values for the interactions in a model, comparing the simulations with a given data set. This lets the researcher check whether the model topology can be a plausible explanation of the reference wet lab data. In case no parameter set can satisfyingly match the given data, or if only a very narrow parameter choice fits well, it is likely necessary to try a different wiring of the network model.

2.6.4. Optimizing model

Our next step was to use ANIMO's automatic parameter fitter on the model, using the wet lab data as reference. We divided the model in two subnetworks roughly corresponding to

Interaction	k*	Qualitative parameter	Scenario	k**
ACAN --I ACAN	0.001	Very slow	1	0.001
AKT --> GSK3	0.008	Fast	1	0.01699021
AKT --I AKT	0.004	Medium	1	0.05987754
AXIN2 --I AXIN2	0.002	Slow	1	0.00016882
Beta-catenin --> TCF/LEF	0.008	Fast	1	0.008
Beta-catenin --I beta-catenin	0.002	Slow	1	0.002
Beta-catenin --I NF-κB	0.008	Fast	1	0.03027397
Beta-catenin --I SOX9	0.016	Very fast	1	0.01824899
BMP2 --> BMPRI/BMPRII	0.016	Very fast	1	0.01769909
BMPR internalization --I BMPRI/ BMPRII	0.016	Very fast	2	0.01426827
BMPRI/BMPRII --> BMPR internalization	0.001	Very slow	1	0.00067129
BMPRI/BMPRII --> SMAD1/5/8	0.008	Fast	1	0.00804472
COL1 --I COL1	0.002	Slow	1	0.002
COL2A --I COL2A	0.001	Very slow	1	0.00042541
COLX --I COLX	0.002	Slow	1	0.002
ERK1/ERK2 --> RUNX2	0.004	Medium	1	0.00417643
ERK1/ERK2 --I ERK1/ERK2	0.004	Medium	1	0.0112296
GSK3 --> beta-catenin	0.016	Very fast	1	0.01599999
GSK3 --> GSK3 phosphatase	0.008	Fast	1	0.02126481
GSK3 phosphatase --I GSK3	0.016	Very fast	2	0.03690672
ID1 --I ID1	0.002	Slow	1	0.00127961
IkB --I NF-κB	0.002	Slow	1	0.00223597
IL-1B --> IL-1RI	0.008	Fast	1	0.01240816
IL1B --I IL-1B	0.002	Slow	1	0.00060578
IL-1BR internalization --I IL-1RI	0.032	Very fast +	2	0.03583533
IL-1RI --> AKT	0.004	Medium	1	0.04684802
IL-1RI --> beta-catenin	0.008	Fast	1	0.00908979
IL-1RI --> ERK1/2	0.008	Fast	1	0.0041156
IL-1RI --> IL-1BR internalization	0.004	Medium	1	0.00782616
IL-1RI --> JNK	0.008	Fast	1	0.016
IL-1RI --> NF-κB	0.016	Very fast	1	0.04698054
IL- 1RI --> p38	0.008	Fast	1	0.04107936
JNK --> RUNX2	0.004	Medium	1	0.00414343

Interaction	k^*	Qualitative parameter	Scenario	k^{**}
JNK --I JNK	0.004	Medium	1	0.00055435
MMP13 --I MMP13	0.002	Slow	1	0.002
NF-κB --> IkB	0.016	Very fast	1	0.016
NF-κB --> IL-1B	0.002	Slow	1	0.00156559
NF-κB --> MMP13	0.001	Very slow	1	0.001
p38 --> RUNX2	0.004	Medium	1	0.00399137
p38 --I p38	0.004	Medium	1	0.04260553
RUNX2 --> COL1	0.002	Slow	1	0.002
RUNX2 --> COLX	0.002	Slow	1	0.002
RUNX2 --> MMP13	0.001	Very slow	1	0.001
RUNX2 --I JNK	0.004	Medium	1	0.00606149
RUNX2 --I p38	0.004	Medium	1	0.00406615
RUNX2 --I RUNX2	0.004	Medium	1	0.01181898
RUNX2 --I SOX9	0.004	Medium	1	0.004
SMAD1/5/8 --> ID1	0.002	Slow	1	0.00270426
SMAD1/5/8 --> RUNX2	0.008	Fast	1	0.008
SMAD1/5/8 --> SMAD7	0.004	Medium	1	0.00671529
SMAD1/5/8 --> SOX9	0.004	Medium	1	0.004
SMAD7 --I SMAD1/5/8	0.008	Fast	2	0.02130379
SOX9 --> ACAN	0.001	Very slow	1	0.001
SOX9 --> COL2A	0.001	Very slow	1	0.001
SOX9 --I beta-catenin	0.001	Very slow	1	0.00028855
SOX9 --I RUNX2	0.004	Medium	1	8.7016E-05
SOX9 --I SOX9	0.002	Slow	1	0.00424899
SOX9-activating signal --> SOX9	0.016	Very fast	1	0.00255911
TCF/LEF --> AXIN2	0.001	Very slow	1	0.00021457
TCF/LEF --I TCF/LEF	0.004	Medium	1	0.004
WNT --> WNTR	0.008	Fast	1	0.02304019
WNTR --> AKT	0.016	Very fast	1	0.04626554
WNTR --> GSK3	0.008	Fast	1	0.00372738
WNTR --> WNTR internalization	0.004	Medium	1	0.00600663
WNTR internalization --I WNTR	0.016	Very fast	1	0.06497972

Table 1. Parameters for WNT, BMP, and IL-1β signaling in the initial model (k^*) and the fitted model that was optimized using experimental data (k^{**}; see Section 2.6.4).

http://dx.doi.org/10.5772/intechopen.69804

the WNT and IL-1 pathways. The two subnetworks were independently fit, based on the wet lab data for the treatments with WNT and IL-1. Dividing the model allowed us to limit the parameter space for the automatic search, making it more rapid: it took less than 3 minutes to complete. In practice, to fit a part of the ANIMO model, we disabled the part of the network we were not focusing on. We then selected those interactions whose parameters we wanted to optimize and clicked on the "Optimize *k*-Values" command in ANIMO's interface. After providing the proper file with the wet lab data, we let the tool to automatically try different *k*-values for the interactions, comparing the model simulations with the data and determining the fitness. Once the tool could find a better fitting set of parameters, the process would terminate, showing the resulting match for the candidate parameter set. We repeated the same procedure on both WNT and IL-1 pathways, fitting the signal transduction parts to the data. We then simulated the WNT + IL-1 treatment in the model and compared it to the data, finding it was fitting already well enough (see **Figure 4**, first and last row).

For BMP2 signaling we optimized SMAD activity based on phosphorylation of Western blot of SMAD1/5/8. SMAD1/5/8 was most active at 15 minutes posttreatment (data not shown).

Finally, we compared the model with the polymerase chain reaction (PCR) experimental data, repeating the fitting process only on that part of the network. The resulting model parameters can be found in **Table 1** (k**), while the comparison between PCR data and ANIMO mRNA node activities is shown in **Figure 5**. We note that the general trends were captured in the model. Differences between activities in the model vs. the experimental data are especially visible in the longer time scales, around 24 hours. This can be expected because higher order effects, such as feedback loops, which take place on longer time scales, are not included in our model.

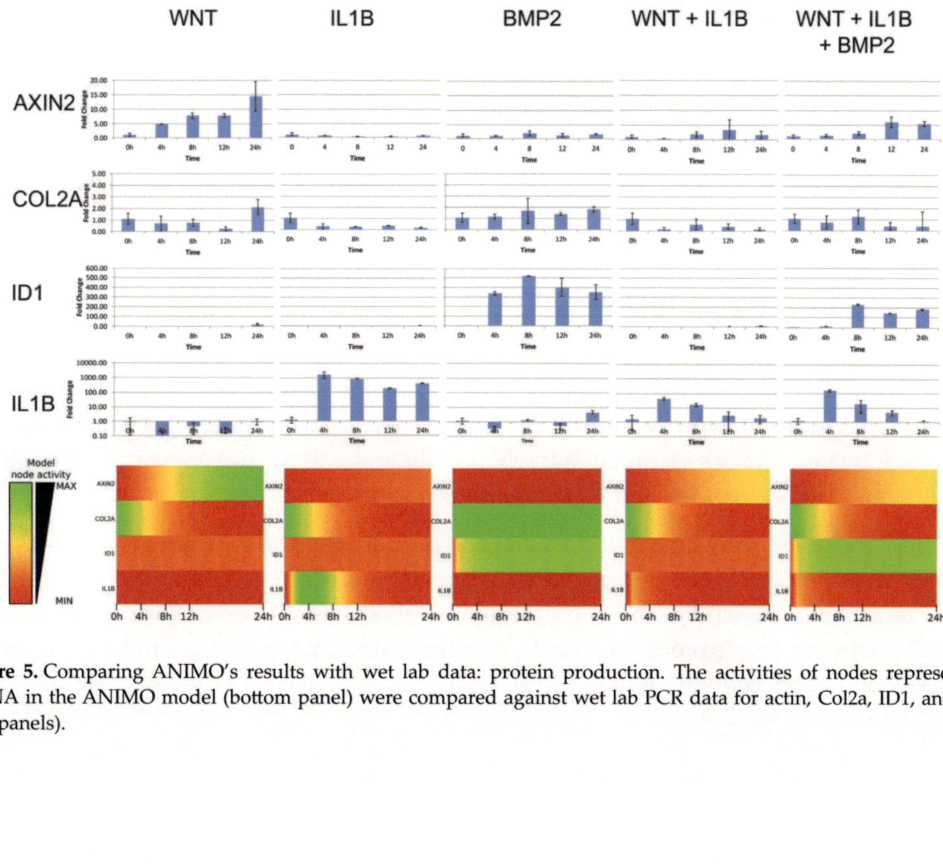

Figure 5. Comparing ANIMO's results with wet lab data: protein production. The activities of nodes representing mRNA in the ANIMO model (bottom panel) were compared against wet lab PCR data for actin, Col2a, ID1, and IL-1 (top panels).

2.6.5. Validating hypothesis

Our hypothesis that IL-1β will increase expression of hypertrophic genes by upregulating RUNX2 activity and downregulating SOX9 activity can now be tested in our optimized model. For this, we can investigate, in silico, the changes in activity of SOX9 and RUNX2 and their corresponding genes, even when no gene expression is measured in the wet lab.

Our model predicts that in the presence of IL-1β and WNT, SOX9 activity will be inhibited and RUNX2 will be activated with an initial peak activity in the first hour (**Figure 6**). This indicates loss of cartilage homeostasis and a slight increase in hypertrophy, which is sustained in time due to the permanent increase in RUNX2 activity (not shown). So even though the initial WNT and IL-1 signal are no longer present, an increase in RUNX2 activity results in a sustained expression of collagen 1 and collagen 10 as well as MMP13, albeit at a low level. Even high levels of BMP2 cannot prevent the loss of SOX9 activity, eventually leading to hypertrophy. These data validate our hypothesis, at least in silico.

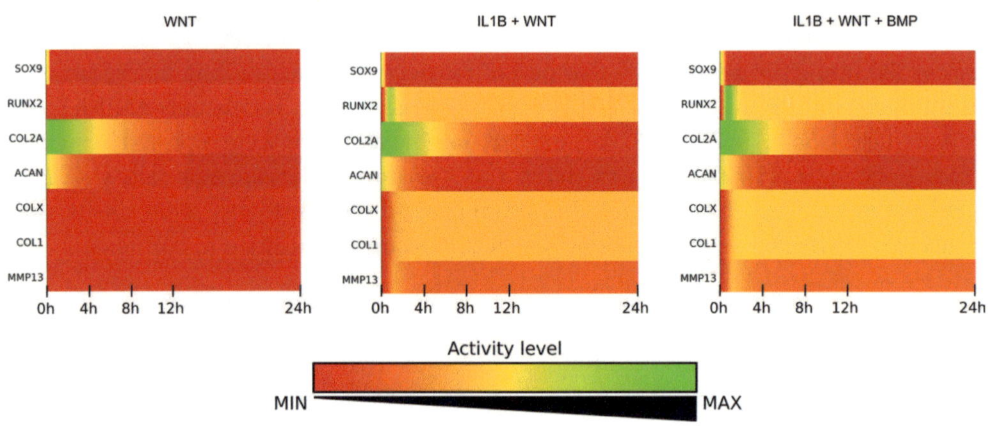

Figure 6. In silico validation of hypothesis: IL-1β will increase expression of hypertrophic genes, *COL1*, *COL10*, and *MMP13*, by upregulating RUNX2 activity and downregulating SOX9 activity.

3. Discussion and conclusion

In this chapter we provide an example of a workflow for starting computational modeling based on literature and experimental data. The aim was not to make the most comprehensive model in terms of network topology, but to understand the dynamics of the network activity in terms of signaling cross talk and corresponding downstream effects.

We chose to use the software ANIMO as a plug-in in Cytoscape as it offers a user-friendly interface in which biologists can interactively create and explore computational models of signal transduction networks. This allows to gain intellectual control over the dynamic behavior of the network that is modeled. We showed that network topologies can be constructed, modified, and enhanced with a formal description of the associated dynamic behavior. The process of modeling biological network dynamics is a prerequisite for formally comparing experimental data to a priori knowledge. ANIMO can also be used in research groups to assist in the storage and transfer of knowledge on biological networks and as a guide in discussions.

Most of the plug-ins for Cytoscape are based on static analyses, for example, they make it possible to find the hubs in a network, to cluster nodes by specific features, or to associate external data sources to the network. This allows to effectively represent large quantities of information and obtain useful insight from them, but the focus is still on the "static picture." ANIMO concentrates instead on the network dynamics: applying an abstract representation of biochemical kinetics, it allows to represent how signaling networks evolve with time under different conditions. Graphs and node colors provide the user with useful representations of the network dynamics. Further analyses are enabled by the possibility to perform model checking on the underlying timed automata model, which can be used without the need to acquire additional training in formal methods.

ANIMO describes biological entities in the network in terms of their activity. This generalizes easily into most signal transduction processes. However, this can also be used to model any process that can be abstractly modeled as a variable activity. Examples are the inclusion of processes such as receptor internalization and phosphatase activity but also inhibition of an activated protein by proteosomal degradation or nuclear export. This flexibility helps the user in describing parts of the network for which the molecular details are unknown or of lesser importance.

In the model presented here, we show how a priori knowledge network based on three signaling pathways can be constructed and tested in silico by asking questions in small steps at a time. We then showed how experimental data of a limited number of proteins and genes, at a wide range of time points, aid to optimize topology and dynamics of the proteins and mRNAs in our network. In the next step, we can prioritize and design new experiments that can be validated in the wet lab. Seeing the role of a computational model in the empirical spiral in **Figure 1**, the work is never finished, but each step in the cycle aids to optimize the model and hence the molecular insight into the dynamics and topology of the cellular signal transduction network.

In ANIMO we proved our hypothesis that IL-1β will increase expression of hypertrophic genes by upregulating RUNX2 activity and downregulating SOX9 activity. For this we used a combination of literature and experimental data to optimize the model parameters. This allowed us to obtain insight into the order of events in the presence of WNT and/or IL-1β at the level of SOX9 and RUNX2 activity. In addition, it allowed insight into the complex interconnectivity of three individual pathways. Such models also yield high content data at high temporal resolution, a feat that is difficult to achieve using only wet lab approaches.

Interestingly, in one computational model, we are able to show a combination of protein activity (phosphorylation) and subsequent mRNA expression. This is a combination model of events at very different time lines. The advantage of our strategy, which included automatic parameter fitting, is the possibility to predict cell fate based on both changes in phosphorylation/protein activity and corresponding gene expression differences.

In the future, ANIMO and related tools may lead to a new paradigm for interactive representation of biological networks. Networks in digital textbooks and articles could be displayed as animations amenable to modifications by readers. Repositories of formal descriptions of signaling modules could be used to put together executable signaling networks. A more user-friendly way of interacting with dynamic network models will lead to a more thorough understanding of biological networks and will accelerate hypothesis-driven research.

Acknowledgements

We thank Patrick Kaufhold and Dr. Jacqueline Plass for the technical assistance. We thank Ricardo Urquidi Camacho for the technical assistance and valuable discussion during the time that he was an MSc graduate student in our lab in 2012 and 2013. We thank Kannan Govindaraj for critically reading the manuscript and for the valuable discussion.

Author details

Stefano Schivo[1], Jeroen Leijten[2], Marcel Karperien[2] and Janine N. Post[2*]

*Address all correspondence to: j.n.post@utwente.nl

1 Formal Methods and Tools, CTIT, University of Twente, Enschede, The Netherlands

2 Developmental BioEngineering, MIRA, University of Twente, Enschede, The Netherlands

References

[1] Hunter T. Protein kinases and phosphatases: The yin and yang of protein phosphorylation and signaling. Cell. 1995;**80**(2):225-236

[2] Cohen P. Targeting protein kinases for the development of anti-inflammatory drugs. Current Opinion in Cell Biology. 2009;**21**(2):317-324

[3] Graves JD, Krebs EG. Protein phosphorylation and signal transduction. Pharmacology and Therapeutics. 1999;**82**(2-3):111-121

[4] Fischer EH. Cellular regulation by protein phosphorylation: A historical overview. Biofactors. 1997;**6**(3):367-374

[5] Pearson RB, Kemp BE. Protein kinase phosphorylation site sequences and consensus specificity motifs: Tabulations. Methods in Enzymology. 1991;**200**:62-81

[6] Mendoza L, Thieffry D, Alvarez-Buylla ER. Genetic control of flower morphogenesis in *Arabidopsis thaliana*: A logical analysis. Bioinformatics. 1999;**15**(7-8):593-606

[7] Shmulevich I, Zhang W. Binary analysis and optimization-based normalization of gene expression data. Bioinformatics. 2002;**18**(4):555-565

[8] de Jong H. Modeling and simulation of genetic regulatory systems: A literature review. Journal of Computational Biology. 2002;**9**(1):67-103

[9] Efroni S, Harel D, Cohen IR. Toward rigorous comprehension of biological complexity: Modeling, execution, and visualization of thymic T-cell maturation. Genome Research. 2003;**13**(11):2485-2497

[10] Fisher J, Piterman N, Hubbard EJ, Stern MJ, Harel D. Computational insights into *Caenorhabditis elegans* vulval development. Proceedings of the National Academy of Sciences of the United States of America. 2005;**102**(6):1951-1956

[11] Ciocchetta F, Duguid A, Gilmore S, Guerriero ML, Hillston J, editors. The Bio-PEPA tool suite. 2009 Sixth International Conference on the Quantitative Evaluation of Systems; IEEE Computer Society. 2009 13-16 Sept.; 2009

[12] COSBILab. COSBILab. 2012. Available from: http://www.cosbi.eu/index.php/research/cosbi-lab

[13] Bartocci E, Corradini F, Merelli E, Tesei L. Model checking biological oscillators. Electronic Notes in Theoretical Computer Science. 2009;**229**(1):41-58

[14] Batt G, Ben Salah R, Maler O. On timed models of gene networks. In: Raskin J-F, Thiagarajan PS, editors. Formal Modeling and Analysis of Timed Systems: 5th International Conference, FORMATS 2007, Salzburg, Austria, October 3-5, 2007 Proceedings. Berlin, Heidelberg: Springer Berlin Heidelberg; 2007. pp. 38-52

[15] Siebert H, Bockmayr A. Temporal constraints in the logical analysis of regulatory networks. Theoretical Computer Science. 2008;**391**(3):258-275

[16] Bonzanni N, Krepska E, Feenstra KA, Fokkink W, Kielmann T, Bal H, et al. Executing multicellular differentiation: Quantitative predictive modelling of *C. elegans* vulval development. Bioinformatics. 2009;**25**(16):2049-2056

[17] Reisig W. Modeling in Systems Biology. Koch I, Reisig W, Schreiber F, editors. Springer London; 2011

[18] Schivo S, Scholma J, Wanders B, Camacho RAU, van der Vet PE, Karperien M, et al. Modelling biological pathway dynamics with Timed Automata. IEEE 12th International Conference on Bioinformatics & Bioengineering; IEEE Computer Society. 2012:447-453

[19] Shannon P, Markiel A, Ozier O, Baliga NS, Wang JT, Ramage D, et al. Cytoscape: A software environment for integrated models of biomolecular interaction networks. Genome Research. 2003;**13**(11):2498-2504

[20] Schivo S, Scholma J, van der Vet PE, Karperien M, Post JN, van de Pol J, et al. Modelling with ANIMO: Between fuzzy logic and differential equations. BMC Systems Biology. 2016;**10**(1):56

[21] Scholma J, Schivo S, Urquidi Camacho RA, van de Pol J, Karperien M, Post JN. Biological networks 101: Computational modeling for molecular biologists. Gene. 2014;**533**(1):379-384

[22] Schivo S, Scholma J, Karperien M, Post JN, Van De Pol J, Langerak R, editors. Setting parameters for biological models with ANIMO. Electronic Proceedings in Theoretical Computer Science, EPTCS; Open Publishing Association. 2014

[23] Johnson VL, Hunter DJ. The epidemiology of osteoarthritis. Best Practice & Research. Clinical Rheumatology. 2014;**28**(1):5-15

[24] Scholma J, Schivo S, Karperien M, Langerak R, van de Pol J, Post JN. An echo in biology: Validating the executable chondrocyte. Osteoarthritis and Cartilage. 2014;**22**:S157-S157

[25] Ahmed TA, Hincke MT. Strategies for articular cartilage lesion repair and functional restoration. Tissue Engineering. Part B, Reviews. 2009;**3**:305-329

[26] Sarzi-Puttini P, Cimmino MA, Scarpa R, Caporali R, Parazzini F, Zaninelli A, et al. Osteoarthritis: An overview of the disease and its treatment strategies. Seminars in Arthritis and Rheumatism. 2005;**35**(1 Suppl 1):1-10

[27] Leijten JC, Emons J, Sticht C, van Gool S, Decker E, Uitterlinden A, et al. Gremlin 1, frizzled-related protein, and Dkk-1 are key regulators of human articular cartilage homeostasis. Arthritis and Rheumatism. 2012;**64**(10):3302-3312

[28] Ling W, Leijten JC, Post JN, Karperien M. Fibroblast growth factor-1 is a mesenchymal stromal cell secreted factor stimulating proliferation of osteoarthritic chondrocytes. Osteoarthritis and Cartilage. 2013;**21**:S273-S274

[29] Ma B, Landman EB, Miclea RL, Wit JM, Robanus-Maandag EC, Post JN, et al. WNT signaling and cartilage: Of mice and men. Calcified Tissue International. 2013;**92**(5):399-411

[30] Ma B, Zhong L, van Blitterswijk CA, Post JN, Karperien M. T cell factor 4 is a pro-catabolic and apoptotic factor in human articular chondrocytes by potentiating nuclear factor kappaB signaling. The Journal of Biological Chemistry. 2013;**288**(24):17552-17558

[31] Zhong L, Huang X, Karperien M, Post JN. The regulatory role of signaling crosstalk in hypertrophy of MSCs and human articular chondrocytes. International Journal of Molecular Sciences. 2015;**16**(8):19225-19247

[32] Clevers H, Nusse R. Wnt/beta-catenin signaling and disease. Cell. 2012;**149**(6):1192-1205

[33] Wang M, Shen J, Jin H, Im HJ, Sandy J, Chen D. Recent progress in understanding molecular mechanisms of cartilage degeneration during osteoarthritis. Annals of the New York Academy of Sciences. 2011;**1240**:61-69

[34] Thomas RS, Clarke AR, Duance VC, Blain EJ. Effects of Wnt3A and mechanical load on cartilage chondrocyte homeostasis. Arthritis Research and Therapy. 2011;**13**(6):R203

[35] Umlauf D, Frank S, Pap T, Bertrand J. Cartilage biology, pathology, and repair. Cellular and Molecular Life Sciences. 2010;**67**(24):4197-4211

[36] Cheng A, Genever PG. SOX9 determines RUNX2 transactivity by directing intracellular degradation. Journal of Bone and Mineral Research: The Official Journal of the American Society for Bone and Mineral Research. 2010;**25**(12):2680-2689

[37] Mackie EJ, Ahmed YA, Tatarczuch L, Chen KS, Mirams M. Endochondral ossification: How cartilage is converted into bone in the developing skeleton. The International Journal of Biochemistry & Cell Biology. 2008;**40**(1):46-62

[38] Goldring MB, Otero M, Plumb DA, Dragomir C, Favero M, El Hachem K, et al. Roles of inflammatory and anabolic cytokines in cartilage metabolism: Signals and multiple effectors converge upon MMP-13 regulation in osteoarthritis. European Cells & Materials. 2011;**21**:202-220

[39] Li X, Tupper JC, Bannerman DD, Winn RK, Rhodes CJ, Harlan JM. Phosphoinositide 3 kinase mediates Toll-like receptor 4-induced activation of NF-kappa B in endothelial cells. Infection and Immunity. 2003;**71**(8):4414-4420

[40] Sonderegger S, Haslinger P, Sabri A, Leisser C, Otten JV, Fiala C, et al. Wingless (Wnt)-3A induces trophoblast migration and matrix metalloproteinase-2 secretion through canonical Wnt signaling and protein kinase B/AKT activation. Endocrinology. 2010;**151**(1):211-220

[41] van der Kraan PM, Blaney Davidson EN, van den Berg WB. Bone morphogenetic proteins and articular cartilage: To serve and protect or a wolf in sheep clothing's?. Osteoarthritis and Cartilage/OARS, Osteoarthritis Research Society. 2010;**18**(6):735-741

[42] Itasaki N, Hoppler S. Crosstalk between Wnt and bone morphogenic protein signaling: A turbulent relationship. Developmental Dynamics. 2010;**239**(1):16-33

[43] Kobayashi M, Squires GR, Mousa A, Tanzer M, Zukor DJ, Antoniou J, et al. Role of interleukin-1 and tumor necrosis factor alpha in matrix degradation of human osteoarthritic cartilage. Arthritis and Rheumatism. 2005;**52**(1):128-135

[44] Wojdasiewicz P, Poniatowski LA, Szukiewicz D. The role of inflammatory and anti-inflammatory cytokines in the pathogenesis of osteoarthritis. Mediators of Inflammation. 2014;**2014**:561459

[45] Weber A, Wasiliew P, Kracht M. Interleukin-1 (IL-1) pathway. Science Signaling. 2010;**3**(105):cm1

[46] Rigoglou S, Papavassiliou AG. The NF-kappaB signalling pathway in osteoarthritis. The International Journal of Biochemistry & Cell Biology. 2013;**45**(11):2580-2584

[47] Daheshia M, Yao JQ. The interleukin 1beta pathway in the pathogenesis of osteoarthritis. The Journal of Rheumatology. 2008;**35**(12):2306-2312

[48] Ma B, van Blitterswijk CA, Karperien M. A Wnt/beta-catenin negative feedback loop inhibits interleukin-1-induced matrix metalloproteinase expression in human articular chondrocytes. Arthritis and Rheumatism. 2012;**64**(8):2589-2600

[49] Nusse R. The WNT Homepage. 2017. Available from: http://web.stanford.edu/group/nusselab/cgi-bin/wnt/

[50] Ducy P, Zhang R, Geoffroy V, Ridall AL, Karsenty G. Osf2/Cbfa1: A transcriptional activator of osteoblast differentiation. Cell. 1997;**89**(5):747-754

[51] Komori T. Regulation of bone development and extracellular matrix protein genes by RUNX2. Cell and Tissue Research. 2010;**339**(1):189-195

[52] Goldring MB, Tsuchimochi K, Ijiri K. The control of chondrogenesis. Journal of Cellular Biochemistry. 2006;**97**(1):33-44

[53] Zhong L, Huang X, Karperien M, Post JN. Correlation between gene expression and osteoarthritis progression in human. International Journal of Molecular Sciences. 2016;**17**(7):1126

[54] Akiyama H, Lyons JP, Mori-Akiyama Y, Yang X, Zhang R, Zhang Z, et al. Interactions between Sox9 and beta-catenin control chondrocyte differentiation. Genes & Development. 2004;**18**(9):1072-1087

[55] Wu M, Chen G, Li YP. TGF-beta and BMP signaling in osteoblast, skeletal development, and bone formation, homeostasis and disease. Bone Research. 2016;**4**:16009

[56] Wuelling M, Vortkamp A. Transcriptional networks controlling chondrocyte proliferation and differentiation during endochondral ossification. Pediatric Nephrology. 2010;**25**(4):625-631

[57] Smith N, Dong Y, Lian JB, Pratap J, Kingsley PD, van Wijnen AJ, et al. Overlapping expression of Runx1(Cbfa2) and Runx2(Cbfa1) transcription factors supports cooperative induction of skeletal development. Journal of Cellular Physiology. 2005;**203**(1):133-143

Chapter 9

FRET-Based Biosensors: Genetically Encoded Tools to Track Kinase Activity in Living Cells

Florian Sizaire and Marc Tramier

Additional information is available at the end of the chapter

http://dx.doi.org/10.5772/intechopen.71005

Abstract

Fluorescence microscopy is widely used in biology to localize, to track, or to quantify proteins in single cells. However, following particular events in living cells with good spatio-temporal resolution is much more complex. In this context, Forster resonance energy transfer (FRET) biosensors are tools that have been developed to monitor various events such as dimerization, cleavage, elasticity, or the activation state of a protein. In particular, genetically encoded FRET biosensors are strong tools to study mechanisms of activation and activity of a large panel of kinases in living cells. Their principles are based on a conformational change of a genetically encoded probe that modulates the distance between a pair of fluorescent proteins leading to FRET variations. Recent advances in fluorescence microscopy such as fluorescence lifetime imaging microscopy (FLIM) have made the quantification of FRET efficiency easier. This review aims to address the different kinase biosensors that have been developed, how they allow specific tracking of the activity or activation of a kinase, and to give an overview of the future challenging methods to simultaneously track several biosensors in the same system.

Keywords: kinase, biosensor, FRET, multiplex, protein conformation, fluorescence microscopy

1. Introduction

Investigating kinase activity in living cells remains a challenge, and usual methods are limited when one wishes to study cellular dynamic events. For a large panel of kinases, the phosphorylation state of the kinase or its substrate has become the main indicator of its activity [1]. One of the most common methods to study this activity is to perform Western Blot analysis on cell extracts by targeting the phosphorylated kinase residue or the phosphorylated substrate residue with an antibody. However, this semi-quantification of the activity state of the kinase applies only for the whole population of the cells [2]. Thus, the other most frequent way to investigate kinase activity is to perform immunofluorescence by targeting the phosphorylated kinase or the phosphorylated substrate with a fluorescent antibody for microscopy observation. This method allows localization of the proteins phosphorylation state in a single cell. However, these two methods suffer from one major limitation: the inability to track this activation state both in space and in time to track dynamic events in living cells. Indeed, this requires lysing the cells or fixing them and permeabilizing them, which prevents sufficient spatio-temporal resolution to investigate intracellular events [3].

To overcome this limitation, new tools have been developed including Forster resonance energy transfer (FRET) biosensors [4]. FRET is a nonradiative transfer of energy of one donor fluorophore to an acceptor fluorophore and relies on (i) an overlap of the emission spectrum of the donor with the excitation spectrum of the acceptor, (ii) an adequate orientation between the two fluorophores, and (iii) a distance less than 10 nm between the two [5]. This feature has been used to investigate various cellular events such as protein–protein interactions by genetically tagging the two proteins of interest with a donor and an acceptor fluorescent protein [6], the intra-cellular Ca^{2+} signal by using calmodulin biosensor [7], proteases activity where the substrate is flanked by two fluorescent proteins, and the decrease of FRET indicates a cleavage of the protein [8], Rho GTPases for cytoskeleton dynamics [9, 10], and mechanical forces at adherent junctions [11, 12, 13]. The first kinase biosensor has been developed for cAMP-dependant protein kinase A (PKA) [14]. From this example and by taking advantage of the FRET characteristics and the conformational modifications of the phosphorylated sensors, several tools to monitor the kinase activity in space and time in living cells have been developed.

By conception, these tools are genetically encoded providing an invaluable advantage to endogenously producing the biosensor in live samples. In this review, we will first present genetically encoded FRET biosensors to monitor kinase activities based on phosphorylated peptide substrate. We will then introduce an alternative way of designing biosensors based on a conformational change of the kinase itself. Finally, we will present new methodological challenges such as multiplex FRET measurements in the same cell, thus allowing simultaneous monitoring of several kinase activities in time and space.

2. Substrate-based kinase activity biosensors

The first genetically encoded FRET biosensor for kinase activity was called A-kinase activity reporter (AKAR) and was designed to investigate the activity of PKA [14]. The idea was to follow a conformational change by FRET in a fusion protein composed of a substrate peptide sequence and a phosphorylated recognition domain. In this foundational work, the biosensor was composed of two fluorescent proteins, CFP and YFP (cyan and yellow fluorescent proteins). Between them sits a first domain, "the peptide substrate," containing a sequence phosphorylated by PKA, followed by a second domain, "the phosphorylation recognition domain," that binds to the peptide substrate when phosphorylated, these two domains are separated by an elastic linker. In the presence of active PKA, the peptide substrate becomes phosphorylated, triggering its affinity for the phosphorylation-binding domain. This association between the two domains induces a conformation change of the biosensor that brings closer both fluorophores and increases FRET efficiency between CPF donor and YFP acceptor (**Figure 1**). The efficiency of FRET can be detected by ratiometric measurements between the intensity signal of the donor and the acceptor. This biosensor can be expressed in cell and is able to provide a response to cell treatment such as forskolin that raises the level of cAMP-activating PKA [14]. This tool has then been improved several times by using a better reversible phospho-binding domain called FHA1 [15] or by changing the fluorophore couple to improve the ratiometric measurements [16, 17]. The AKAR biosensor has been used to report the activity of PKA in neurons of mouse brain slices, showing its value in neurosciences [18].

Based on this concept, several new kinase FRET biosensors were developed. The E-kinase activity reporter (EKAR) biosensor is a FRET-based probe to study ERK activity [19]. The fluorophore pair is composed of the green donor eGFP and the red acceptor mRFP1. The consensus substrate peptide originates from Cdc25c, a member of the MAPK family. As other kinases from this family could phosphorylate the substrate, an ERK binding domain has been inserted to ensure ERK specificity. A WW domain (containing 2 tryptophans separated by around 20 aa) was used to bind the phosphorylated substrate [20], and a flexible linker allows a conformational change, when the Cdc25C peptide substrate is phosphorylated. This biosensor has also experienced several steps of optimization by modifying the fluorophores or the flexible linker [21, 22]. Among all these biosensors, the reversibility of the conformational modification is a major feature to study variations of kinase activation states [23].

Kinase biosensors are such powerful tools to investigate the dynamic of kinase activity events in cells that several of these biosensors have been created to study mitotic kinases activity through the cell cycle, including cyclin B1-Cdk1 [24]. A kinase biosensor has also been used to study PKC (protein kinase C) activation which is involved in tumor promotion. CKAR (C-kinase activity reporter) is composed of the CFP/YFP fluorophore pair, a specific peptide substrate for PKC, and the FHA2 domain of Rad53p that can bind to the phosphorylated substrate [25]. In this particular case, the unphosphorylated biosensor harbors a maximum

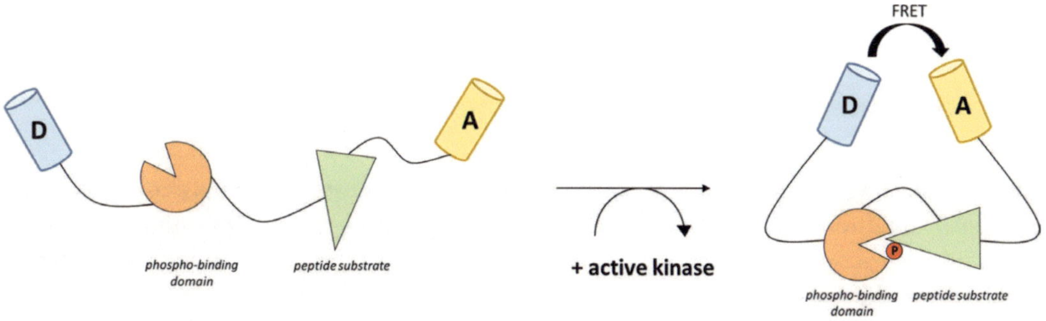

Figure 1. Mechanism of a substrate-based FRET biosensor. When the biosensor is not phosphorylated, it adopts an opened conformation keeping away the donor fluorophore D, and the acceptor fluorophore, A. After the peptide substrate phosphorylation by the kinase, a phospho-binding domain can bind to it gathering the fluorophore pair and allowing FRET.

FRET efficiency conformation, and FRET signal decreases once phosphorylated. By rapidly acquiring FRET efficiency, oscillations of PKC phosphorylation in a range of a minute were highlighted [25].

Other biosensors have been derived from CKAR. Polo-like kinase-1 (PLK1) is a major mitotic kinase that activates Cdc25C phosphatase, which abrogates the inhibitory phosphorylation of proteins controlling the entry to mitosis. A FRET-based biosensor has been created by replacing the peptide substrate of PKC with a peptide substrate of PLK1, the use of which revealed that the timed-control activation of PLK1 depends on Aurora A [26]. The choice of the kinase peptide substrate to construct the biosensor is a key point to improve its specificity. For Plk1, a c-Jun substrate-based biosensor was developed [27], since the previous version based on Myt1 substrate sequence was also sensitive to Mts1 activity [28]. The c-jun–based version was then used to demonstrate that Plk1 activity is required for commitment to mitosis during cell cycles [29].

A biosensor to study Aurora B activation has also been developed [30]. But in this work, authors wanted to monitor the kinase activity at a specific location, since it has been postulated that an activity gradient of Aurora B at the mitotic spindle may play a role for mitotic progression. If one considers a conventional version of the FRET biosensor, its diffusion throughout the cell is too fast, and it is not possible to reveal a precise localization of the activation. Thus, Fuller and co-workers have added different localization sequences to target the biosensor either to the centromere using a peptide from CENP-B or to the chromatin using histone H2B [30].

FRET-based substrate kinase biosensors are good tools to investigate kinase activity, but they have some limitations and present three major challenges: (i) the biosensor relies on the endogenous kinase phosphorylating the substrate peptide, and thus, FRET variation is observed only when the kinase is particularly abundant or heavily stimulated, (ii) the sequence flanking the phosphorylation residue(s) targeted by the kinase must be known and selective for the kinase under study, and (iii) these biosensors only explore the catalytic activity of the kinase toward a specific substrate at once and not the activation process of the kinase itself. To solve this last issue, a new set of kinase FRET biosensors has been developed based on conformational changes of the kinase when active.

3. Conformational kinase-based biosensor

An alternative way of genetically encoded FRET-based substrate kinase biosensors has been developed by directly using the full-length kinase peptide sequence. Activation of a kinase frequently relies on a conformational opening of the enzymatic pocket. The idea is then to tag the whole kinase at its N- and C-terminus with a FRET pair of fluorescent proteins to be able to monitor this kinase activation related to the conformational change (**Figure 2**). To our knowledge, the first kinase FRET biosensor using this concept was developed to study c-Raf conformation [31]. This biosensor called Prin-c-Raf uses the CFP/YFP pair to flank c-Raf. A flexible linker has been added between the acceptor fluorophore and the kinase to enhance

http://dx.doi.org/10.5772/intechopen.71005

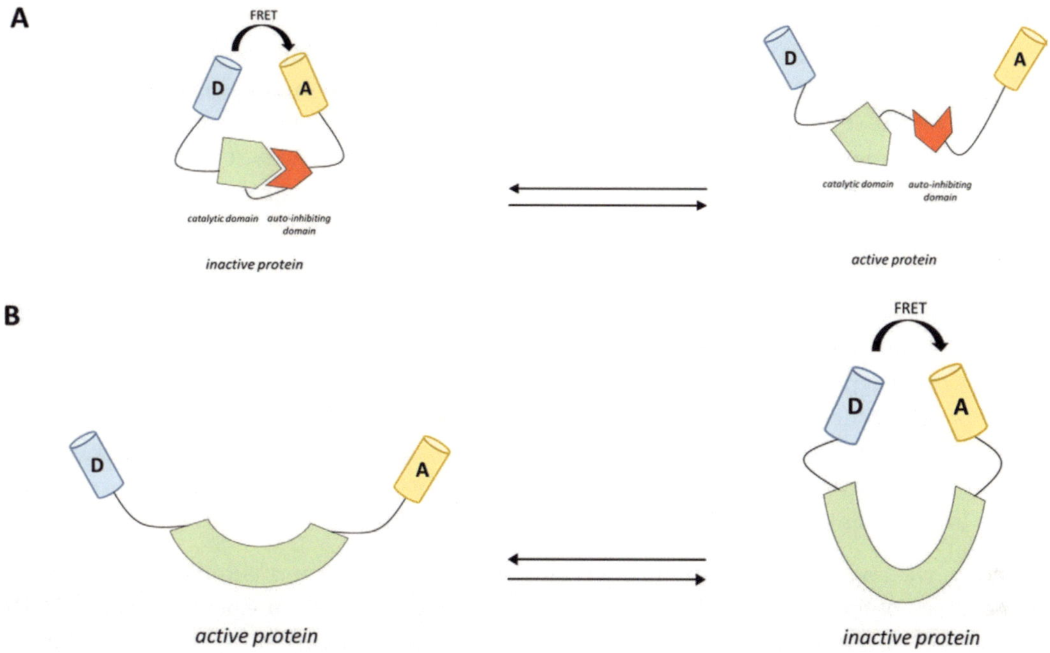

Figure 2. Mechanisms of a conformational-based FRET biosensor. (A) The auto-inhibiting domain can bin the catalytic domain of the kinase bringing closer the donor fluorophore D and the acceptor fluorophore A, allowing FRET. When the kinase is activated, the auto-inhibiting domain unbinds and the kinase adopts an opened and active conformation with a FRET decrease. (B) A lot of proteins adopt a new conformation when activated that modulates the distance between the pair of fluorophores.

FRET efficiency. Mutation of the residues Ser259 and Ser261 preventing c-RAF phosphorylation and mimicking the active state of the kinase leads to an open conformation of Prin-c-RAF as FRET ratio is decreased. When a constitutively active mutant of AKT that negatively regulates c-RAF is expressed, the wild-type version of c-RAF shows high FRET signal consecutive to a closed inactive conformation, while the mutated version S259A and S261A stays open with a lower FRET ratio. By using this biosensor, authors were able to show that the constitutively active H-RasV12 localized at the plasma membrane binds and opens the wild-type biosensor in an active conformation, inducing the recruitment of MEK at the plasma membrane.

A biosensor for PKCγ consisting of the kinase flanked by the donor super cyan fluorescent protein 3 (SCFP3A) and the acceptor YFP has also been developed [32]. The kinase displays a pseudosubstrate domain that is able to bind and inhibit catalytic activity. In this work, they compared a different mutated form of the biosensor PKCγ-A24E, where the pseudosubstrate cannot bind to PKCγ and observed a decrease of FRET that they are associated with an opened active conformation.

Another conformational biosensor has been developed to study FAK (Focal Adhesion Kinase) activity by taking advantages of the conformational changes associated with the activation state of the kinase being controlled by an inhibitory domain [33]. For that, a biosensor was constructed with the full-length kinase containing a FERM domain (F for 4.1

protein, E for ezrin, R for radixin, and M for moesin) for membrane localization and a kinase domain using the CFP/YFP FRET pair, the donor at the N-terminus of the protein and the acceptor directly between the two domains. When the FERM domain binds to the catalytic domain of FAK, it inhibits the kinase activity and FRET occurs. On the contrary, the absence of FRET corresponds to an active and thus open conformation. It is then possible to monitor FAK activity at the focal adhesion of living cells by expressing the biosensor transiently in living cells.

A biosensor of maternal embryonic leucine zipper kinase (MELK) has been created consisting of the MELK sequence flanked by the CFP/YFP pair [34]. As well as FAK, MELK has an auto-inhibited domain at the C-terminus that can bind to the catalytic domain of the kinase. This biosensor was expressed in *Xenopus* embryos, and conformational changes were monitored in dividing cells. It has been demonstrated that the biosensor exhibits a closed conformation in the cytosol and an open conformation at the cleavage furrow. But here again, as for previous conformational sensors, only conclusions on the conformational change of the kinase could be made. Its direct link to the kinase activation (and activity) was not tackled.

Recently, we have developed an Aurora A biosensor based on conformational changes [35]. It is composed of the full-length kinase flanked by a GFP donor and a mCherry acceptor. To be functional, Aurora A undergoes a conformational change following autophosphorylation on the T288 residue [36, 37]. By exploiting this mechanism, we designed a biosensor that directly associates the conformational change of the kinase with its state of activation. Indeed, *in vitro* treatment with ATP leads to a closed conformation when treatment with phosphatase leads to an opened conformation. The activation state is also monitored by fluorescence lifetime imaging microscopy (FLIM) in living cells. Through this work, we show that the biosensor was able to functionally replace the endogenous Aurora A depleted by siRNA. Thus, by replacing the endogenous kinase, this biosensor is a direct reporter of the activation state of Aurora A at endogenous levels in stable cell lines with a good spatio-temporal resolution. With this tool, by dissociating the quantity and the activation state of the kinase, we were able to highlight a new nonmitotic role of Aurora A in G1 phase [35].

This kind of biosensor can also be adapted to other enzymatic activities. As an example, a BRET (bioluminescence resonance energy transfer)-based biosensor of the PTEN (phosphatase and tensin homolog) phosphatase has been developed and is composed of the full-length PTEN protein flanked by a donor Rluc and the acceptor YFP [38]. PTEN biosensor immunoprecipitated from cells displays the same phosphatase activity on PIP3 (phosphatidylinositol 3,4,5 trisphosphate) and AKT (or Protein kinase B) as the wild-type PTEN. This biosensor can also be expressed at endogenous levels in human embryonic kidney (HEK) cells. The mutation of four residues Ser380, Thr382, Thr383, and Ser385 favoring a closed conformation leads to a strong decrease of BRET signal. The association between conformational changes and the activity state has allowed the monitoring of PTEN regulation in living cells. Authors have thus been able to correlate inhibition of the known activation pathway of PTEN using a CK2 (Casein Kinase 2) inhibitor, with its change of conformation or inversely by co-expressing S1PR2, an activator of PTEN. Once the biosensor was validated, the authors used it to identify new GPCRs (G protein–coupled receptors) activating PTEN.

The reliability of these biosensors, consisting of the full-length protein flanked by a pair of fluorophores, was recently applied to study any protein function associated with a conformational change. For example, a study of the conformational change of the Tau protein has been tackled using the protein flanked by a CFP/YFP pair [39]. The use of this biosensor led to the demonstration that the binding of Tau to the microtubules induces a switch to a hairpin conformation of Tau. In addition, it has been shown that mutations of Tau responsible of Frontotemporal dementia with parkinsonism-17 (FTDP-17) disorder alter this conformation change. This method has also been used to study vinculin conformation by using a biosensor consisting of the protein flanked by an mTurquoise donor and a NeonGreen acceptor [40]. Vinculin displays an auto-inhibited state, when the tail domain and the head domain are binding, increasing FRET signal. A mutated version of the biosensor that is unable to bind to talin showed a decreasing FRET signal and a disruption in the vinculin localization at focal adhesions. In contrast, paxillin knock-down or mutations leading to a decrease in actin binding did not modify FRET signal.

Thus, these genetically encoded biosensors are efficient at monitoring protein activation at cellular levels when expressed in living cells. A lot of proteins are known to adopt different conformation states according to their activation, and this is why FRET or BRET biosensors are best suited for tracking their activity in space and in time in living cells. It is likely that this tool will be used intensively to study protein conformation linked to activity in the next few years.

One can thus follow the activation of a kinase by following its conformational change using a conformation-based biosensor and follow its catalytic activity using a substrate-based biosensor. It would be of great interest to simultaneously follow activation and activity in a single living cell, a pursuit that calls for methods able to monitor the two different FRET biosensors simultaneously.

4. New methodological insights for multiplexing kinase biosensors

Owing to complex crosstalk between signaling pathways, multi-parameter biosensing experiments have become essential to correlate biochemical activities without lag time during a dedicated cellular process. A very exciting challenge has thus been to follow several FRET biosensors on the same sample at the same time and in the same location [41]. Commonly, FRET is measured by the fluorescence intensity ratio of the acceptor to the donor. In that case, whatever the two fluorescent protein FRET pairs chosen, CFP/YFP and mOrange/mCherry [42], mTFP1/mCitrine and mAmetrine/tdTomato [43, 44], mTagBFP/sfGFP and mVenus/mKok [45], the multiplex approach suffers from two limitations: (i) a spectral bleed-through of the first acceptor in the second donor emission band that depends directly on the respective quantities of the two biosensors and (ii) the multiple excitation wavelength which requires sequential acquisition that does not adequately follow fast signal dynamics or signal changes in highly motile samples.

To overcome the first limitation, a meroCBD (merocyanine–Cdc42-binding domain) biosensor modified with a far-red organic fluorophore (Alexa750) was used for probing Cdc42 simultaneously with a genetically encoded CFP/YFP FRET-based biosensor for Rho A [46]. This approach prevents spectral bleed-through but cannot be generalized to all genetically encoded FRET biosensors, where organic fluorophores cannot easily replace fluorescent proteins. The same team also developed an environment-sensing dye called mero199 [47]. This dye can bind to the active form of Cdc42 leading to a shift of its excitation/emission ratio. In combination with a Rac1 biosensor, they were able to simultaneously monitor activation of both proteins and to correlate it with retraction or velocity of migrating MEF (mouse embryonic fibroblasts) cells. Very recently, an elegant method based on linear unmixing of 3D excitation/emission fingerprints applied to three biosensors simultaneously was published [48]. This type of approach based on image calculation is often limited by the different biosensors expression levels and a poor signal-to-noise ratio after complex image corrections.

To overcome the second limitation, the two FRET pairs CFP/YFP and Sapphire/RFP in combination with a single violet excitation were used [49], resulting in no lag time in biochemical activity recording. But again, in this case, the spectral bleed-through and excitation crosstalk necessitate linear unmixing. Another interesting approach for simultaneously multiplexing two FRET activities was developed using a "Large Stokes Shift" orange fluorescent protein, LSSmOrange [50]. The authors used a CFP-YFP together with LSSmOrange-mKate2 biosensors enabling imaging of apoptotic activity and calcium fluctuations in real time using intensity-based methods. Other studies were carried out utilizing FLIM instead of ratio imaging to measure FRET. When FRET occurs, donor fluorescence lifetime decreases. This method requires measurement of the donor fluorescence only and is independent of emission from the acceptor. By using CFP and YFP as donor and the same red acceptor (tHcRed), FLIM of CFP and YFP donors allow the two different FRET signals to be distinguished [51]. Combination of FLIM-FRET of a red-shifted TagRFP/mPlum pair with ratio imaging of a CFP/Venus pair allows maximal the spectral separation while, at the same time, overcoming the low quantum yield of the far-red acceptor mPlum [52]. The two last examples alleviated the spectral bleed-through but not the limitation associated with multiple excitations.

To overcome both limitations, a novel red-shifted fluorophore mCyRFP1 has been developed with a high Stokes shift [53]. This fluorophore has an excitation spectrum in the range of the GFP emission spectrum (around 500 nm), but its emission spectrum is shifted compared to GFP. An emission dichroic filter allows simultaneous detection of the GFP fluorescence lifetime and the mCyRFP1 fluorescence lifetime. The authors were able to perform two-photon fluorescence lifetime imaging by using only one excitation laser at 920 nm with a Rhoa biosensor and a CaMKIIα biosensor. Furthermore, while the RhoA biosensor uses the pair mCyRFP1/mMaroon1, the CaMKIIα biosensor uses mEGFP and dimVenus which is a dark fluorophore preventing bleed-through with mCyRFP1.

Recently, our team has developed a similar method by taking advantages of the LSSmOrange (Large Stoke Shift) and the dark fluorophore ShadowG [54]. We modified two substrate kinase biosensors, EKAR2G and AKAR4 (E-Kinase Activity Reporter type 2G for ERK and A-Kinase Activity Reporter type 4 for PKA), with a new pair of fluorophores mTFP1/ShadowG and LSSmOrange/mKate2, respectively. LSSmOrange and mTFP1 are both excitable by using a single 440 nm wavelength. By single excitation wavelength dual-color FLIM, we are able to simultaneously monitor the activity of ERK and PKA in living cells at the same location. Thus, the activity of each kinase in response to forskolin or EGF treatment can be imaged simultaneously. This approach overcomes the limitations of the multiple excitation wavelengths and bleed-through.

Because FLIM is now a widely used microscopy approach, the decrease of the donor lifetime is sufficient to quantify FRET, and fluorescence of the acceptor is not mandatory, as it is still the case when one uses ratiometric FRET. Changing the fluorescent acceptor with a nonfluorescent acceptor leads to the development of a new kind of single-color FRET biosensor (**Figure 3A**). It is then perfectly adapted for simultaneous monitoring of kinase FRET biosensing.

Another method to get a single-color biosensor to perform multiplex could be based on homoFRET measured by anisotropy [55, 56]. HomoFRET occurs when a fluorophore transfers its energy to a closely identical fluorophore. However, it is impossible to measure homoFRET by ratiometric or fluorescence lifetime measurements. Fluorescence anisotropy can be measured by detecting the parallel and the perpendicular light emitted by a fluorophore excited with a polarized light [57], and this anisotropy decreases when FRET occurs between nonparallel fluorescent dipoles. This method was already used to study protein oligomerization [58]. For example, a study has used fluorescence anisotropy to determine the degree of clustering of proteins such as GPI or EGFR fused to GFP in living cells [59]. This approach has been investigated to multiplex at the same time a conventional calcium heteroFRET biosensor using FLIM with the oligomerization of pleckstrin homology domains of Akt (Akt-PH) labeled

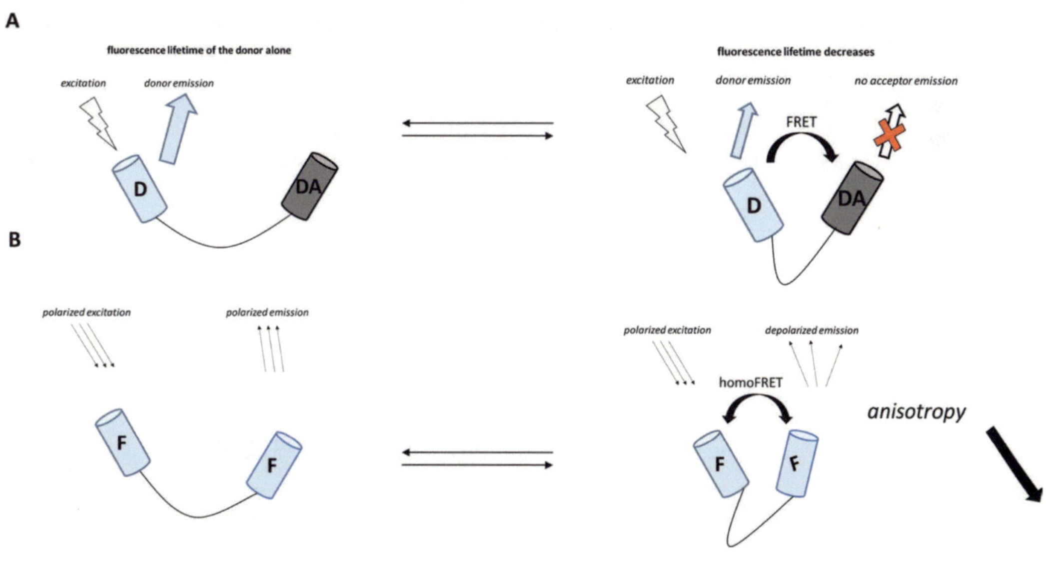

Figure 3. Single-color genetically encoded FRET biosensor. (A) When a fluorophore F is excited, it can transfer its energy to a dark acceptor DA by FRET. Even excited, the dark acceptor emits no detectable light, and FRET is measured by the measurement of donor fluorescence lifetime. It constitutes a single-color FRET biosensor. (B) When a fluorophore F is excited by a polarized light, it emits polarized fluorescence. When homoFRET occurs between two identical fluorophores, it can leads to the depolarization of fluorescence emission decreasing the anisotropy. Again, it constitutes a single-color FRET biosensor.

with mCherry [60]. From our knowledge, the development of an intramolecular homoFRET biosensor to follow a biochemical activity was not yet developed but has very interesting potential. Adapting this method to kinase biosensors would provide a new methodology to simultaneously follow multiple biosensors.

5. Concluding remarks

Kinases have multiple functions in cells, and their mechanisms are very dynamic in both space and time. We have focused our review on two types of kinase biosensors. The substrate-based kinase biosensors are good tools to specifically monitor the activity of a kinase, but it requires to have a good knowledge of the substrate peptide sequence, particularly for its specificity, and a biological system where the activity of the kinase is sufficient to detect FRET. The conformation-based biosensors provide information about the activation state of the kinase itself; however, they do not provide information about its catalytic activity that can be further regulated by other post-translational modifications. Gathering these different tools with a multiplex methodology by using the approaches of single-color FRET biosensor would provide new mechanistic insight to investigate kinase functions with an adequate spatio-temporal resolution.

Acknowledgements

We wish to apologize to the authors whose works were not cited in this review. We wish to thank G. Bertolin, G. Herbomel, and C. Demeautis for their many helpful discussions about FRET biosensors and D. Fairbass for critical reading the manuscript. Work in the laboratory is supported by the CNRS and the University of Rennes 1 and grant from the "Ligue Nationale Contre le Cancer" (LNCC, region Grand Ouest). F. S. is fellow of the Région Bretagne and University of Rennes 1.

Author details

Florian Sizaire and Marc Tramier*

*Address all correspondence to: marc.tramier@univ-rennes1.fr

Institut de Génétique et Développement de Rennes (IGDR), CNRS, Université de Rennes, Rennes, France

References

[1] Manning G, Whyte DB, Martinez H, Sudarsanam S. The protein kinase complement of the human genome. Science. 2002;**298**:1912-1934

[2] Yang P-C, Mahmood T. Western blot: Technique, theory, and trouble shooting. North American Journal of Medical Sciences. 2012;**4**:429

[3] Koley D, Bard AJ. Triton X-100 concentration effects on membrane permeability of a single HeLa cell by scanning electrochemical microscopy (SECM). Proceedings of the National Academy of Sciences. 2010;**107**:16783-16787

[4] Aoki K, Kamioka Y, Matsuda M. Fluorescence resonance energy transfer imaging of cell signaling from in vitro to in vivo : Basis of biosensor construction, live imaging, and image processing. Development, Growth & Differentiation. 2013;**55**:515-522

[5] Padilla-Parra S, Tramier M. FRET microscopy in the living cell: Different approaches, strengths and weaknesses. BioEssays. 2012;**34**:369-376

[6] Kenworthy AK. Imaging protein-protein interactions using fluorescence resonance energy transfer microscopy. Methods. 2001;**24**:289-296

[7] Miyawaki A, Llopis J, Heim R, McCaffery JM, and others (1997). Fluorescent indicators for Ca^{2+} based on green fluorescent proteins and calmodulin. Nature 388, 882.

[8] Horton RA, Strachan EA, Vogel KW, Riddle SM. A substrate for deubiquitinating enzymes based on time-resolved fluorescence resonance energy transfer between terbium and yellow fluorescent protein. Analytical Biochemistry. 2007;**360**:138-143

[9] Itoh RE, Kurokawa K, Ohba Y, Yoshizaki H, Mochizuki N, Matsuda M. Activation of Rac and Cdc42 video imaged by fluorescent resonance energy transfer-based single-molecule probes in the membrane of living cells. Molecular and Cellular Biology. 2002;**22**:6582-6591

[10] Hinde E, Digman MA, Hahn KM, Gratton E. Millisecond spatiotemporal dynamics of FRET biosensors by the pair correlation function and the phasor approach to FLIM. Proceedings of the National Academy of Sciences. 2013;**110**:135-140

[11] Herbomel G, Hatte G, Roul J, Padilla-Parra S, Tassan J-P, Tramier M. Actomyosin-generated tension on cadherin is similar between dividing and non-dividing epithelial cells in early *Xenopus laevis* embryos. Scientific Reports. 2017;**7**:45058

[12] Borghi N, Sorokina M, Shcherbakova OG, Weis WI, Pruitt BL, Nelson WJ, Dunn AR. E-cadherin is under constitutive actomyosin-generated tension that is increased at cell-cell contacts upon externally applied stretch. Proceedings of the National Academy of Sciences. 2012;**109**:12568-12573

[13] Grashoff C, Hoffman BD, Brenner MD, Zhou R, Parsons M, Yang MT, McLean MA, Sligar SG, Chen CS, Ha T, et al. Measuring mechanical tension across vinculin reveals regulation of focal adhesion dynamics. Nature. 2010;**466**:263-266

[14] Zhang J, Ma Y, Taylor SS, Tsien RY. Genetically encoded reporters of protein kinase A activity reveal impact of substrate tethering. Proceedings of the National Academy of Sciences. 2001;**98**:14997-15002

[15] Zhang J, Hupfeld CJ, Taylor SS, Olefsky JM, Tsien RY. Insulin disrupts β-adrenergic signalling to protein kinase A in adipocytes. Nature. 2005;**437**:569-573

[16] Allen MD, Zhang J. Subcellular dynamics of protein kinase A activity visualized by FRET-based reporters. Biochemical and Biophysical Research Communications. 2006;**348**:716-721

[17] Depry C, Allen MD, Zhang J. Visualization of PKA activity in plasma membrane microdomains. Molecular BioSystems. 2011;**7**:52-58

[18] Gervasi N, Hepp R, Tricoire L, Zhang J, Lambolez B, Paupardin-Tritsch D, Vincent P. Dynamics of protein kinase A signaling at the membrane, in the cytosol, and in the nucleus of neurons in mouse brain slices. Journal of Neuroscience. 2007;**27**:2744-2750

[19] Harvey CD, Ehrhardt AG, Cellurale C, Zhong H, Yasuda R, Davis RJ, Svoboda K. A genetically encoded fluorescent sensor of ERK activity. Proceedings of the National Academy of Sciences. 2008;**105**:19264-19269

[20] Lu P-J, Zhou XZ, Shen M, Lu KP. Function of WW domains as phosphoserine-or phosphothreonine-binding modules. Science. 1999;**283**:1325-1328

[21] Vandame P, Spriet C, Riquet F, Trinel D, Cailliau-Maggio K, Bodart J-F. Optimization of ERK activity biosensors for both ratiometric and lifetime FRET measurements. Sensors. 2014;**14**:1140-1154

[22] Komatsu N, Aoki K, Yamada M, Yukinaga H, Fujita Y, Kamioka Y, Matsuda M. Development of an optimized backbone of FRET biosensors for kinases and GTPases. Molecular Biology of the Cell. 2011;**22**:4647-4656

[23] González-Vera J, Morris M. Fluorescent reporters and biosensors for probing the dynamic behavior of protein kinases. Proteomes. 2015;**3**:369-410

[24] Gavet O, Pines J. Activation of cyclin B1–Cdk1 synchronizes events in the nucleus and the cytoplasm at mitosis. The Journal of Cell Biology. 2010;**189**:247-259

[25] Violin JD, Zhang J, Tsien RY, Newton AC. A genetically encoded fluorescent reporter reveals oscillatory phosphorylation by protein kinase C. The Journal of Cell Biology. 2003;**161**:899-909

[26] Macůrek L, Lindqvist A, Lim D, Lampson MA, Klompmaker R, Freire R, Clouin C, Taylor SS, Yaffe MB, Medema RH. Polo-like kinase-1 is activated by aurora A to promote checkpoint recovery. Nature. 2008;**455**:119-123

[27] Liu D, Davydenko O, Lampson MA. Polo-like kinase-1 regulates kinetochore–microtubule dynamics and spindle checkpoint silencing. The Journal of Cell Biology. 2012;**198**:491-499

[28] Bruinsma W, Macůrek L, Freire R, Lindqvist A, Medema RH. Bora and aurora-A continue to activate Plk1 in mitosis. Journal of Cell Science. 2014;**127**:801-811

[29] Gheghiani L, Loew D, Lombard B, Mansfeld J, Gavet O. PLK1 activation in late G2 sets up commitment to mitosis. Cell Reports. 2017;**19**:2060-2073

[30] Fuller BG, Lampson MA, Foley EA, Rosasco-Nitcher S, Le KV, Tobelmann P, Brautigan DL, Stukenberg PT, Kapoor TM. Midzone activation of aurora B in anaphase produces an intracellular phosphorylation gradient. Nature. 2008;**453**:1132-1136

[31] Terai K, Matsuda M. Ras binding opens c-Raf to expose the docking site for mitogen-activated protein kinase kinase. EMBO Reports. 2005;**6**:251-255

[32] Verbeek DS, Goedhart J, Bruinsma L, Sinke RJ, Reits EA. PKC mutations in spinocerebellar ataxia type 14 affect C1 domain accessibility and kinase activity leading to aberrant MAPK signaling. Journal of Cell Science. 2008;**121**:2339-2349

[33] Cai X, Lietha D, Ceccarelli DF, Karginov AV, Rajfur Z, Jacobson K, Hahn KM, Eck MJ, Schaller MD. Spatial and temporal regulation of focal adhesion kinase activity in living cells. Molecular and Cellular Biology. 2008;**28**:201-214

[34] Le Page Y, Chartrain I, Badouel C, Tassan J-P. A functional analysis of MELK in cell division reveals a transition in the mode of cytokinesis during Xenopus development. Journal of Cell Science. 2011;**124**:958-968

[35] Bertolin G, Sizaire F, Herbomel G, Reboutier D, Prigent C, Tramier M. A FRET biosensor reveals spatiotemporal activation and functions of aurora kinase A in living cells. Nature Communications. 2016;**7**:12674

[36] Walter AO, Seghezzi W, Korver W, Sheung J, Lees E. The mitotic serine/threonine kinase Aurora2/AIK is regulated by phosphorylation and degradation. Oncogene. 2000;**19**(42):4906-4916

[37] Bayliss R, Sardon T, Vernos I, Conti E. Structural basis of Aurora-A activation by TPX2 at the mitotic spindle. Molecular Cell. 2003;**12**(4):851-862

[38] Lima-Fernandes E, Misticone S, Boularan C, Paradis JS, Enslen H, Roux PP, Bouvier M, Baillie GS, Marullo S, Scott MGH. A biosensor to monitor dynamic regulation and function of tumour suppressor PTEN in living cells. Nature Communications. 2014;**5**

[39] Di Primio C, Quercioli V, Siano G, Rovere M, Kovacech B, Novak M, Cattaneo A. The distance between N and C termini of tau and of FTDP-17 mutants is modulated by microtubule interactions in living cells. Frontiers in Molecular Neuroscience. 2017;**10**

[40] Case LB, Baird MA, Shtengel G, Campbell SL, Hess HF, Davidson MW, Waterman CM. Molecular mechanism of vinculin activation and nanoscale spatial organization in focal adhesions. Nature Cell Biology. 2015;**17**:880-892

[41] Carlson HJ, Campbell RE. Genetically encoded FRET-based biosensors for multiparameter fluorescence imaging. Current Opinion in Biotechnology. 2009;**20**:19-27

[42] Piljic A, Schultz C. Simultaneous recording of multiple cellular events by FRET. ACS Chemical Biology. 2008;**3**:156-160

[43] Ai H, Hazelwood KL, Davidson MW, Campbell RE. Fluorescent protein FRET pairs for ratiometric imaging of dual biosensors. Nature Methods. 2008;**5**:401-403

[44] Ding Y, Ai H, Hoi H, Campbell RE. Förster resonance energy transfer-based biosensors for multiparameter ratiometric imaging of Ca^{2+} dynamics and caspase-3 activity in single cells. Analytical Chemistry. 2011;**83**:9687-9693

[45] Su T, Pan S, Luo Q, Zhang Z. Monitoring of dual bio-molecular events using FRET biosensors based on mTagBFP/sfGFP and mVenus/mKOκ fluorescent protein pairs. Biosensors & Bioelectronics. 2013;**46**:97-101

[46] Machacek M, Hodgson L, Welch C, Elliott H, Pertz O, Nalbant P, Abell A, Johnson GL, Hahn KM, Danuser G. Coordination of Rho GTPase activities during cell protrusion. Nature. 2009;**461**:99-103

[47] MacNevin CJ, Gremyachinskiy D, Hsu C-W, Li L, Rougie M, Davis TT, Hahn KM. Environment-sensing merocyanine dyes for live cell imaging applications. Bioconjugate Chemistry. 2013;**24**:215-223

[48] Woehler A. Simultaneous quantitative live cell imaging of multiple FRET-based biosensors. PLoS One. 2013;**8**:e61096

[49] Niino Y, Hotta K, Oka K. Simultaneous live cell imaging using dual FRET sensors with a single excitation light. PLoS One. 2009;**4**:e6036

[50] Shcherbakova DM, Hink MA, Joosen L, Gadella TWJ, Verkhusha VV. An orange fluorescent protein with a large stokes shift for single-excitation multicolor FCCS and FRET imaging. Journal of the American Chemical Society. 2012;**134**:7913-7923

[51] Peyker A, Rocks O, Bastiaens PIH. Imaging activation of two Ras isoforms simultaneously in a single cell. Chembiochem. 2005;**6**:78-85

[52] Grant DM, Zhang W, McGhee EJ, Bunney TD, Talbot CB, Kumar S, Munro I, Dunsby C, Neil MAA, Katan M, et al. Multiplexed FRET to image multiple signaling events in live cells. Biophysical Journal. 2008;**95**:L69-L71

[53] Laviv T, Kim BB, Chu J, Lam AJ, Lin MZ, Yasuda R. Simultaneous dual-color fluorescence lifetime imaging with novel red-shifted fluorescent proteins. Nature Methods. 2016;**13**:989-992

[54] Demeautis C, Sipieter F, Roul J, Chapuis C, Padilla-Parra S, Riquet FB, Tramier M. Multiplexing PKA and ERK1&2 kinases FRET biosensors in living cells using single excitation wavelength dual colour FLIM. Scientific Reports. 2017;**7**:41026

[55] Gautier I, Tramier M, Durieux C, Coppey J, Pansu RB, Nicolas J-C, Kemnitz K, Coppey-Moisan M. Homo-FRET microscopy in living cells to measure monomer-dimer transition of GFP-tagged proteins. Biophysical Journal. 2001;**80**:3000-3008

[56] Tramier M, and Coppey-Moisan M. (2008). Fluorescence anisotropy imaging microscopy for homo-FRET in living cells. Methods in Cell Biology, (Elsevier), pp. 395-414.

[57] Chan FTS, Kaminski CF, Kaminski Schierle GS. HomoFRET fluorescence anisotropy imaging as a tool to study molecular self-assembly in live cells. Chemphyschem. 2011; **12**:500-509

[58] Yeow EKL, Clayton AHA. Enumeration of oligomerization states of membrane proteins in living cells by homo-FRET spectroscopy and microscopy: Theory and application. Biophysical Journal. 2007;**92**:3098-3104

[59] Bader AN, Hofman EG, Voortman J, van Bergen en Henegouwen PMP, Gerritsen HC. Homo-FRET imaging enables quantification of protein cluster sizes with subcellular resolution. Biophysical Journal. 2009;**97**:2613-2622

[60] Warren S, Margineanu A, Katan M, Dunsby C, French P. Homo-FRET based biosensors and their application to multiplexed imaging of signalling events in live cells. International Journal of Molecular Sciences. 2015;**16**:14695-14716